Natural Goat Care

Natural Goat Care

Pat Coleby

Acres U.S.A.
Austin, Texas

Natural Goat Care

Copyright © 2001, 2012 by Pat Coleby

All rights reserved. No part of this book may be used or reproduced without written permission except in cases of brief quotations embodied in articles and books.

The information in this book is true and complete to the best of our knowledge. All recommendations are made without guarantee on the part of the author and Acres U.S.A. The author and publisher disclaim any liability in connection with the use or misuse of this information.

Acres U.S.A.
P.O. Box 301209
Austin, Texas 78703 U.S.A.
(512) 892-4400 • fax (512) 892-4448
info@acresusa.com • www.acresusa.com

Printed in the United States of America.
Printed on recycled paper.

Publisher's Cataloging-in-Publication

Coleby, Pat, 1928-
Natural goat care / Pat Coleby, Austin, TX, ACRES U.S.A., 2012
xii, 372 p., 23 cm.
Includes bibliographical references and index.
Library of Congress catalog card number: 00-101379
ISBN: 978-0-911311-66-2 (trade)

1. Goats. 2. Goats — Nutrition — Requirements.
3. Goats — Health. 4. Goats — Breeding. I. Coleby, Pat, 1928-,
II. Title.

SF383.C65 2012   636.39'08

*Dedicated to goats of all kinds.
There are no other animals quite like them.*

Acknowledgements

Thanks to Owen and the late Iris Dawson for checking the early manuscript, and to Owen for providing many of the illustrations. Sandy Green, Wallace Keir, Noreen Hicks, Owen Dawson and many more provided photos which were gratefully received. Also thanks to Sarah Gundry and family of Cohuna Cashmeres for pictures of Cashmeres and Cashgoras. I am grateful to Dr. Bernard Jensen for permission to quote the milk table in Chapter 13. And finally, my thanks to the many goats that I have farmed over the years and who have taught me so much.

Contents

Acknowledgments vii

Preface xi

Chapter 1. A Brief Look at Goats Worldwide 1

Chapter 2. Land, Control, Housing
& Farming Methods 13

Chapter 3. Different Kinds of Husbandry 33

Chapter 4. Acquiring Stock 45

Chapter 5. Breeds 67

Chapter 6. Nutritional Requirements
& Basic Feeding Practices 89

Chapter 7. Psychological Needs of Goats 115

Chapter 8. Management 121

Chapter 9. Minerals: Their Uses
& Deficiency Signs 153

Chapter 10. Vitamins & the Use of Herbal,
Homeopathic & Natural Remedies. 177

Chapter 11. Health Problems 193

Chapter 12. Breeding & Selection
for Desirable Characteristics 299

Chapter 13. Goats for Milk. 313

Chapter 14. Goats for Meat and Skins 337

Chapter 15. Showing Goats 345

Bibliography 359

Index 365

Preface

I wrote *Australian Goat Husbandry* in the late seventies, those heady days when none of us had heard of Caprine Arthritis Encephalitis virsus (CAE) or AIDS-related diseases. Since then we have learned much — not all of it pleasant — about the farming and care of goats. Every time I check some information in the first book I realize how far we have come in the last 20 years.

Natural Goat Care came out in 1993, but was three years in the writing. It now goes into its second and updated edition in Australia, and I am proud that this is the first edition of the book to be published by Acres U.S.A. in America.

The intent of this book is to help goat farmers of all kinds by writing an easily understood and comprehensive book on all aspects of goat farming and looking after goats, whatever the numbers — two or two hundred — or kind: fleece, meat, dairy or hobby (we do not have pygmy goats yet in Australia). There is a great deal of highly technical information available and I have tried to present it in simple terms. Those who are interested can read on for further information. The book should be studied at the outset of any operation, and not just kept until an animal is ill and then referred to for treatment options.

Goats are animals that above all thrive on fully organic natural conditions on the farm; this book tells how it can be done.

Much of the husbandry is the direct result of experience, other people's and my own, over a period of many years; like everyone I am still learning — the most recent victory was an answer to blackleg that worked. All the husbandry consists of non-invasive remedies that are as natural as possible combined with basic land care on the Albrecht model.

Owen and the late Iris Dawson of Pearcedale, Victoria helped me enormously by checking the manuscript of the first edition of *Natural Goat Care* and pointing out my mistakes, and Owen (who is still very much with us) did most of the illustrations. His encyclopedic knowledge of goats and how they are made has always been invaluable to the goat-keeping fraternity.

Pat Coleby, 2000
Maldon, Victoria, Australia

Chapter 1

A Brief Look at Goats Worldwide

Goats, the "poor man's cow," were possibly the first animal to be domesticated for man's use. Certainly their origins as companions and servants of the human race are lost in the mists of time. Yet, paradoxically, they remain

Boer and feral goats.

quite the most undomesticated of our animals and will easily revert to their feral habits. The United States has its indigenous goats unlike Australia where they were brought over with the first fleet and successive ships.

For Feeding Children

From the time of the goat's first arrival in Australia, children have been reared on goat's milk when their natural source of milk was not functioning. The does either suckled the children directly, or a form of "teat," which was usually a piece of rag dipped in the milk for the child to suck, was used until the invention of the teat as we know it today.

Disease-free Status

Goat's milk was always highly regarded because goats had the reputation (not always deserved) of being much healthier than other animals. Unfortunately we know now that goats are subject to diseases like all stock, but, if well kept, definitely do something to uphold their reputation as a disease-free animal.

Travellers

Travellers on ships before the days of refrigeration took goats with them to ensure a fresh milk (and meat) supply, and marvelously, the animals survived. Voyagers in Australia's First Fleet, as is shown by the following excerpt from Captain John Hunter's *Voyage* (Page 31), took their goats along.

> *Table Bay (South Africa) October 14th 1787. We embarked on board Sirius . . . with a number of sheep, goats, hogs and poultry . . . . A quantity of livestock also put on the store ships amounted to . . . four goats. The officers each provided themselves with livestock . . . not merely for the voyage, but with a view to stocking their little farms in the country to which they were going; every person in the fleet determined to live wholly on salt provisions in order to land as much livestock as possible.*

The latter was sacrifice indeed and showed how seriously the possession of some source of milk was regarded. In more recent times the German Warship Emden, which was captured in World War I, also carried goats for the milk supply for at least some of the crew. The descendants of these blue-eyed Saanens were still to be found in the 1950s in Australia.

Goats Around the World

Until our knowledge of the cause and spread of disease brought about modern quarantine laws, goats were shipped all over the world from their various countries of origin. Australia and New Zealand were probably the only countries that did not have indigenous goats of some kind.

In New Zealand, goats were evidently left by Captain Cook's original fleet; the ones on the mainland may have disappeared but he also landed some on Arapawa Island. The descendants of these were not examined closely until 1973, when the goats were found to be quite unlike any breed in the world today (see Chapter 12).

Milking Goats

Most of our milking goats come from the mountains of Europe, the well known Saanens, Toggenburgs and French Alpines (from which derive the British Alpines). The oldest book I have on goats was published in Chicago by the American Sheep Breeders Association in 1903. It is *A Manual of Angora Goat Raising* with a chapter on "Milch Goats," by George Fayette Thompson, who had already written several books on Angora goats. At that time the difference between Cashmere and Angora goats was not fully understood by everyone. The book traces the history of nearly all the goats then known in the Northern hemisphere.

Nubians, as we know them, were the result of crossing some of the Eastern breeds and in the 19th century were considered the best milking goats in the world, famous for their long lactations and high production (Fayette

French Alpine kid.

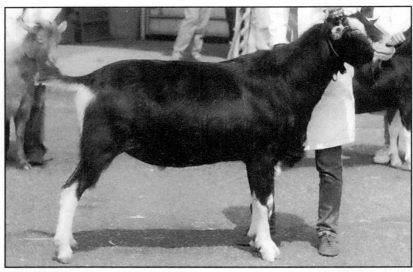

British Alpine buck.

Toggenburg buck.

Thompson). Europe, Asia, Africa and, in particular, Spain, Malta, India, China, the Americas and the Middle East all have their varieties of goat, but apparently there were not many serious attempts to breed them up for milk as was done with the Swiss and Nubian breeds.

Nubian doe.

Goats for Fleece

Fleece goats also did their share of travelling around the world, probably as a source of meat on board the sailing ships, as well as the hoped for nuclei of Angora or Cashmere flocks in the new countries. Angoras are believed to have evolved from goats in Persia known as the Paseng or Bezoar, but opinions differ as to whether *Capra falconeri* or *Capra aegagrus* was the actual progenitor of the Angoras as we know them today. Both types were famous for long, spirally-twisted horns and very fine silky fleeces. Hair from goats has been spun since pre-biblical times for making curtains, tents and temple furniture. Their skins were used for clothes and containers for liquids by nomadic peoples worldwide. Possibly we would not recognize those goats today as either Angoras or Cashmeres — perhaps something in between.

Cashmere and Angora Origins

Angora (from whence mohair goats get the name), in what used to be known as Asia Minor and is now Turkey, is about 250 miles from Constantinople. It is a harsh mountainous region where the winter temperatures are extremely cold, swinging up to 35 to 40 degrees celsius (95 degrees Fahrenheit

Angora goats.

and above) in the summer, rainfall being about eight inches per annum on average.

Photographs taken in Turkey during the 1870s show small, well-covered animals with no facial cover, but the fleeces are obviously fine and of good character. It was noted by travellers who managed to battle their way into the remote region that all animals in the area, dogs and cats as well as goats, had exceptionally fine, silky hair. This was put down to climate and/or some other factor and many believed that this would be lost when the goats were taken to other regions. However, the importations of these goats to America, after initial troubles which were mainly due to the goats that were used for upgrading (a common cause of difficulty), were found by 1900 to be producing fleeces that compared favorably with those of the country of origin.

The grease in the fiber of the Texan goats, and to a degree the South Africans, was apparently a characteristic of many early Angoras in their native land. The Texans have, for various reasons, selected for it and it is now an accepted characteristic of those goats. Australian Angora breeders ended up with a greaseless goat, which has not helped the shearing process, but has spared them the trauma of fly struck animals. Texas Angoras, like Merino sheep can and do get struck. This can of course be avoided with good husbandry.

Types of Angora fleece vary; this probably stems back to the beginning of the 19th century when the purebred goats of Angora could not provide enough fleece to meet the demand. Upgrading was the only answer — as it has been many times since — Kurdish goats providing the numbers. This resulted in a mixture of types that still shows up nearly two centuries later. A very bad drought in the latter part of the nineteenth century killed many of the goats of Angora. Again, other breeds were brought in to build up the stocks. General consensus reckoned that the goat that resulted was a better animal in many respects than its predecessor. The bucks were reported as producing fleeces of about six to nine pounds and the does fleeces of about one to three pounds.

In the 1850s Turkey (Asia Minor) was producing about two million pounds of mohair per annum. It was reported at that time that the mohair from South Africa, in spite of some very fine specimens of Angora goat in that country, was of inferior quality.

First Fiber Goats in the United States

In the mid-nineteenth century a Dr. Davis, who was sent to Turkey from the United States to help in the culture of cotton, so impressed the Sultan that he presented Davis with a flock of Angoras among which was one purebred Tibet goat — a Cashmere — as a reward. When these goats came to their new home in the United States, they passed into the hands of a Colonel Peters and for many years, even as late as 1860, were all regarded as Cashmeres. They were thought to be the goats from whose fleeces the famous Paisley shawls were made, these spun in the Paisley district of northern England. They were spun so finely from the cashmere that it was said they could be passed through a woman's wedding ring.

Cashmere goats.

Eventually the mix-up in identities was sorted out and the two breeds whose product, except for the fineness of the fleece when compared with sheep's wool, is not so very similar, were accepted as breeds in their own right. Most of the Angoras were farmed in the southern states of America and, incredibly, appear to have survived the carnage of the Civil War to found what has been one of the great Angora populations of the 20th century. There were at least three more importations into North America of Angoras during the latter part of the 19th century and the earlier part of the 20th century, all of which helped establish the foundation for the importations to Australia in the 1980s and 1990s.

Mohair Production

Since those days the farming and breeding of Angoras in the United States, particularly Texas, has been responsible (along with South Africa) for much of the mohair processed in the world today. The first Texan Angoras were released from quarantine in Australia in 1992. The South Africans followed soon after. Mohair figures from the 1988 International Mohair Association (I.M.A.) report are: Turkey 5.6 million pounds, South Africa 23 million pounds, Texas 14 million pounds, Australia two million pounds, Argentina, Lesotho and New Zealand were still below the two million pound mark.

Cashgora buck. This breed is a cross between the Angora and Cashmere.

Fiber Goats in Australia

From 1960 onward farmers in Australia have worked on improving the quality and quantity of both mohair and cashmere so by the latter part of the 20th century the production is a force in world markets for both fibers. *Cashgora*, the cross between the two fleece breeds, is also becoming a marketable commodity; the New Zealanders in particular have concentrated on this product.

Meat Industry

In Australia, the meat industry has dragged behind the other goat concerns for many years; partly, but not wholly, due to the seasonal and uncertain supply. However, the arrival of the Boers some years ago has changed all that and they are now very much part of the farming scene and are spreading across the world. In Africa they have been known for many years as a docile, easily handled and fast-maturing meat goat. In the last thirty years they have been promoted in a very professional way, and as a result are now to be found in most countries. They come in black and white and there are now an appreciable number of the reds around as well. Broadacre sheep and cattle farmers are taking on these animals as an adjunct to their herds as they do not need extra fencing and are much easier to control than fleece or feral goats.

Commercial Milking

Commercial milking operations are to be found all over the world, running from 20 to 200 goats. The majority of these operations produce cheese and yogurt and a few also sell unprocessed milk. In Europe, commercial milking goat farms are the norm and there are many big farms mostly, but not all, running Saanen goats, especially in Holland. In the United Kingdom, the United States and Australia they have only been an accepted part of the farm scene since the end of World War II.

Pasteurization

Unfortunately in Australia it has not been possible for the small goat keeper to sell unpasteurized milk for some years now, following some very dirty milk that came on the market. Sadly the industry brought the prohibition on itself. In the United Kingdom the rule has been relaxed for specific dairies and hopefully we may yet get back to selling wholesome *clean* milk again in the not too distant future. This will be covered at length in Chapter 13.

Goats on Farms

Goats are at last beginning to be accepted, even by those not connected with them, as productive farm animals for fleece, milk and meat instead of a music hall joke. In Australia, fleece and Boer goats are now being run in conjunction with many of the big Merino sheep flocks. Station owners have realized that goats complement many other grazing animals due to their slightly different eating patterns.

Chapter 2

Land, Control, Housing & Farming Methods

Goats have higher mineral requirements than other domesticated animals because they are natural browsers. Deer and camelids share this characteristic, as those running them in traditional farming situations are discovering. Trees and shrubs are higher in minerals than grass. Their deep roots obtain minerals from way down in soils. These nutrients have not been leached and farmed out of this deep soil as they have on the surface or topsoil.

Those who farm goats or any other browser must remember this high mineral requirement. If they do not, the moment of truth — with sick and dying goats — almost invariably occurs within three years. When the land is as bad as in the analysis provided below, poor animal health arrives much earlier. The deficiencies catch up to the animals and the farmer is left wondering what happened. It is at this point that the enterprise is usually abandoned. Meat and fleece goats that are not invariably hand-fed, as milkers often are, can be given the basic stock lick ingredients used by other species (explained in Chapter 6). This supplement should provide them much of what they cannot obtain from their environment.

Soil Analysis

The single most important factor when considering goat farming is to have a full analysis done of the land so that the exact state of the mineral balance is known and remedial action can be taken where necessary. When the mineral balance is known, the goats' diets may be supplemented with those nutrients that are missing. Treating the land with minerals is less damaging for the environment than conventional fertilizers and goats, like all stock, do not benefit from the latter. In fact, whenever there is trouble on the farm, it can often be traced back to the annual top dressing with chemical fertilizers.

Top dressing will depend entirely on the calcium to magnesium ratios; strangely enough it is possible to have a near perfect pH and for those ratios to be right out of balance. So an analysis of the soil is *essential*. Then the land must be treated with the lime minerals: gypsum, lime and/or dolomite depending on the ratios of sulfur, calcium or magnesium shown. It is important to have the total phosphorus monitored because it would be very unusual for that mineral to have to be applied (this is true in Australia at any rate). I have yet to see an analysis where there is not a reasonably large "bank" of locked up phosphorus; this, according to the agronomist Neal Kinsey, cannot be used until the calcium and magnesium in the soil are at the right level.

Analyses throughout the farming areas in Australia show enormous deficiencies in some or all minerals. Calcium, magnesium and sulfur are always in short supply or badly out of balance. This is partly due to the great age of this continent and millennia of leaching, but also to the overuse of sodium compounds and/or acidifying chemical fertilizers which have been found to inhibit many minerals, including magnesium, sulfur and all of the trace minerals. The latter cannot be spread or replaced while the pH is below 5.5 or they will be leached out, and when the top dressing brings the pH up to that level, many of the trace minerals become available again anyway.

In most districts of Australia, dolomite (which is about 54 percent calcium and 25 percent magnesium, depending on where it is mined) can be spread. There are areas where magnesium is plentiful and this is often the norm in the United States, but it cannot be taken for granted. So an analysis is necessary because in that case only calcium, in the form of agricultural lime, will be needed. Sulfur is another important mineral nearly always in short supply. According to CSIRO (Commonwealth Scientific Industry Research Organization), sulfur is inhibited when conventional fertilizers have been overused. Gypsum, (calcium sulfate) can be used alone or mixed with dolomite or lime to replace it, or, in non-clay soils, yellow sulfur may be mixed in with the top dressing. Top dressed minerals can take a year or more to reach the food chain — longer in a drought or when the soil is particularly poor. But usually a marked improvement in pasture quality will be noted the spring after the first top dressing.

However, in the United States, as in the United Kingdom and New Zealand, the opposite picture may often be found. I work with one commercial goat facility in the United Kingdom and have had to stop them from feeding grain and similar feeds. Their hay and silage is of such high quality that extra protein feeds are not needed. The goats only need to have access to their minerals, preferably on free-choice. At least one of my correspondents in the United States uses this method successfully.

Other land improvement measures which will help restore the humus in the soil should include soil aeration, tree planting programs and spreading manure. This last is better composted if possible, but spreading it straight from the sheds is better than none at all provided the lime levels are monitored.

Feeding Minerals

For milking goats it is necessary to provide supplementary sources of trace minerals and dolomite as an ongoing part of the husbandry. Their requirement is greater than can ever be met by grazing — the amounts of extra minerals may be reduced over time according to how sick the land was originally, but they can never be eliminated. Fleece and meat goats, once the land is in really good heart, will need few extras as the demands on them are not so great as the demands on milkers or breeders. Trace minerals that were originally missing will gradually become available again as the health and balance of the pastures improves.

Pasture and Scrub

Goats thrive on mixed pasture that contains what are commonly known as weeds as well as grasses and legumes. Blackberry, thistles, docks and different kinds of herbs, shrubs, fodder trees — both European and local — all are much prized. However, fodder trees must be protected or the goats will eat them out as they do blackberries and most other weeds over time. The minerals in all these are higher than those found in grass alone and goats "do" very well while clearing them out. Goats that receive copper in their lick offerings and diet will not ring-bark trees, even if they eat the foliage of some of them when young. Trees like oak, ash and many other northern hemisphere trees make good natural browsing for goats.

Meat and feral goats have been used successfully in blackberry and scrub clearing programs. The milkers would have had their udders torn and the fiber goats would have got hung up by their fleeces, but ferals and Boers do very well as scrub clearers. As one forestry works manager put it to me: "Getting the goats in was one of the best moves I ever made, the men hated using the chemicals, but they enjoy having the goats. The clearing is being done more successfully and the morale is better." In that case, it was a reforestation program and about 40 goats of indeterminate ancestry were acquired, with a quantity of electric fencing. The whole operation cost considerably less than using chemicals and manpower and was more effective than the spraying program that preceded it.

Fodder trees should be planted in belts behind fences through which the goats can browse, but not destroy them. Mountainous, rocky and well-treed areas, ideally with some coast line, make the best goat country. David Mackenzie, the Scottish author of the classic book, *Goat Husbandry*, writes of goats kept in those conditions. This is why he can blandly aver that goats should not get diseases; they would be unlikely to do so in such situations as they can obtain everything they need. However, for the majority of goat farmers, wherever they are, their goats have to survive and stay healthy in the farm complex, which is not nearly so easy.

Goats, like all farm animals, are better lightly stocked — 10 goats can be run to the acre, but a great deal of experience and expertise is needed to do it successfully. Two goats to the acre is more realistic; farmers do not realize that the destructiveness of goats is caused by their high mineral requirement. If they have the right supplements, especially adequate copper, they will not willfully eat bark and destroy trees. This applies to all animals; the fact is that the animals will do whatever they can to satisfy their nutritional needs — including devouring trees.

Following is an extract from a mineral analysis done by a very good independent firm in Australia. The United States has several such too and good advisors, for instance, Neal Kinsey to name one. The farm studied below was very sick and the stock was only just holding its own. One of the most important items is the organic matter; if it is plentiful even poor mineral levels can be tolerated for a while, the one shown here is passable.

The "desired level" is an indication of the best levels for mixed pastures under reasonable rainfall conditions. These vary. It takes years to degrade the soil — the restoration can be achieved much faster. The unexpected bonus of this sort of reclamation project is not only the visible improvement of the pasture, but even before that becomes obvious the improvement in the condition of goats is very marked, they do best on organically treated soils.

Soil Analysis

Items	Result	Desirable level
Soil Color: dark gray		
Texture: fine sand loam		
pH (1.5 water)		5.00
pH (1:5 01M c12)		4.50
Electric Conductivity, EC uS/cm	200.00	<300.00
Total Soluble Salt, TSS/ppm	660.00	<300.00
Available Calcium, Ca/ppm	860.00	1,800-2,400
Available Magnesium, Mg/ppm	160.00	430-600
Available Sodium, Na/ppm	207.00	<184.00
Available Hydrogen, H/ppm	114.00	<64.00
Available Nitrogen, N/ppm	8.20	25.00
Available Phosphorus, P/ppm	13.60	30.00
Available Potassium, K/ppm	121.00	190.00
Available Sulfur, S/ppm	4.10	7.00
Available Copper, Cu/ppm	0.20	2.00
Available Zinc, Zn/ppm	3.00	7.00
Available Iron, Fe/ppm	23.00	>20.00
Available Manganese, Mn/ppm	2.00	>20.00
Available Cobalt, Ca/ppm	0.03	>1.00
Available Molybdenum, Mo/ppm	0.30	1.00
Available Boron, B/ppm	0.60	0.6-1.0
Total Organic Matter, OM/%	9.90	6-10.00
Total Phosphorus, P/ppm	2.00	17.00
Cation Exchange Capacity, CEC		16.04
Exch. Sodium Percentage, ESP	4.93	<5.00
Calcium/Magnesium Ratio, Ca/Mg	3.23	2-4.00

Percentages of Ca, Mg, Na, K and percentage of CEC

	Percentage	Percentage of CEC
Exchangeable Calcium, Ca	62.90%	23.57%
Exchangeable Magnesium, Mg	19.47%	7.29%
Exchangeable Sodium, Na	13.14%	4.93%
Exchangeable Potassium, K	4.49%	1.68%
Exchangeable Hydrogen, H	—	62.53%
	100.00%	100.00%

Recommendations

Gypsum	1 Ton to the acre
Dolomite	1 Ton to the acre

The analysis above is a fair example of a good-to-middling analysis in Australia. There will certainly be areas in the United States that are as poor, but as one of the agronomists who writes for *Acres U.S.A.* told me, "We do not expect to farm in those areas." In Australia we have very little usable land, so we have to learn how to manage it — which has certainly taught me a lot.

Once the gypsum and dolomite have been spread, the soil should be well on the way to recovery if the process is followed up by aeration. This applies anywhere in the world where soil compaction is a serious and ongoing problem. In his excellent book, *Weeds: Control Without Poisons*, Charles Walters points out that compaction is nearly always a contributory factor in weed infestations. Low potassium will come up when the land is being treated naturally and until it does, using cider vinegar in the animal's diet will be useful for supplying it (see Chapter 10).

Iodine is not, strictly speaking, a mineral and cannot be shown on an analysis, but most of Australia and Tasmania are iodine deficient. Other countries often have induced iodine shortfalls due to artificial fertilizers and, if there is any doubt about iodine levels, feed seaweed products on demand to supply any shortfall; the goats will know whether they need it or not.

Irrigation

After living in a drought belt for many years where the goats were very healthy, I moved to an irrigation farm with relief, thinking they would enjoy year-round green grass. Five years later I finally admitted that goats are not suited to that type of farming; nor actually are most northern hemisphere varieties of stock. The goats always chose the non-irrigated parts of the farm to graze, even in the drought. Like most grazing animals they tend to do better on dry country if feed is obtainable.

Control

This can mean different things in different countries — Maltese and Greek herdsmen control their goats by age-old methods that do not involve fences, except for yarding them at night. The goats learn that they stay within a certain distance of the herdsmen or a stick that he puts in the ground. Goats can be, as people soon discover, highly intelligent and responsive to human beings.

Herding Australian style.

The easiest way to control goats is to have their pastures attractive enough for them to want to stay on the right side of the fence. Also, as they are gregarious, it is easier to keep several goats in a paddock — one alone may be quite unhappy and determined to escape. When I bring strange goats into my herd situation, beyond a few battles to establish the pecking order, they never try to leave the others. Particular attention should be paid to the prevailing wind. Like sheep, goats go into the wind and the stronger it is the more they do so. For example, in an area where north winds are frequent, all fences on the north side (particularly if they are against a road) must be extremely well built.

Fencing

There is a bewildering choice of fencing on the market these days — many of them practicable for goats. The ideal permanent fence for me is eight-strand cattle boundary netting with two or three plain wires on the top about four to five inches apart. Another strand should be strained along the bottom of the netting, and an optional one placed in the middle. All should be stapled to the netting. This fencing is good for goats who are not starving and/or not jumpers — both are equally difficult to control. Personally I would not have a jumper on the place — goats are very quick learners.

The fences should be good quality ring-lock type, some farmers tell me they prefer hinge joint because the joints do not come apart under pressure like cheap ring-lock. I find goats eventually (sooner rather than later) fold hinge joint like a concertina, no matter how well strained. Bird or rabbit netting is useless for goats, no matter how well supported; they always end up annihilating it.

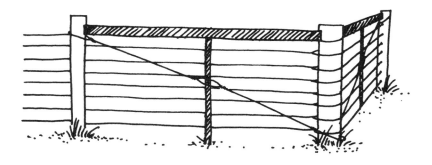

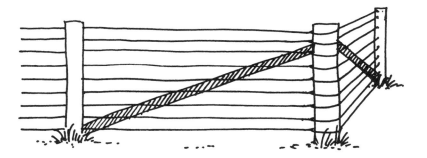

Fencing corner assemblies.

All wire fences must be properly strained and the strainer sections preferably should be the square type. If the one-post type with the angled support, it must be fully enclosed or the goats will run up it.

Buck Runs

If money is unlimited, posts three yards or so apart with four or five good rails or weld-mesh or chain-link fencing are ideal. For the last two, the posts should be a maximum of three yards apart and the wire five to six-feet high. Even on farms where good cattle wire fences are used, runs to contain bucks or young, in-season females must be built like this, nothing else will keep the bucks and the does apart. I have known large, overly keen bucks (milk type) to break through a brand new weld-mesh fence in 24 hours — in these cases reinforcement is obviously necessary. Heavy pipe (old pipe is ideal), one placed two feet from the ground and another a further one-foot-eight-inches above it at butting height works well. Bucks as keen and strong as that are not too usual.

Wooden paling fences (in countries where wood is cheap and easy to get) make good paddock fences for small areas, the palings should be four-inches apart, and four-feet high.

Electric Systems

Electric fences can be used as a back up. The buck mentioned previously would have thought twice if there had been two "hot wires" at butting height inside his run. These strands should be about four to five inches out from the fence and are effective after training — one of my bucks made three attempts at biting them before he admitted defeat.

An electric wire run about eight to 10 inches out from the top of a fence will stop goats from leaning over to reach trees. If a fence is to be totally electrified, the strands will have to be near enough to stop goats from getting through them, i.e., five or six wires. Alternatively electric netting, which is expensive, can be used for small areas. Goats have the ability, many times proven, to assess when an electric fence is *not* functioning because they can hear the current passing through. I ran a single strand at the top of a fence and have often seen them "listening" to find out if it was operative.

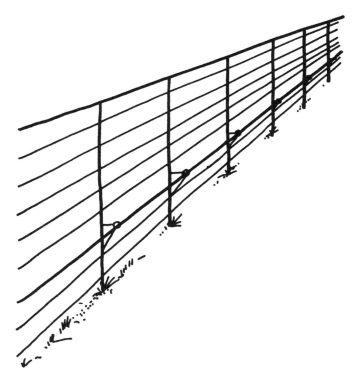

"Hot" wire at butting or rubbing level.

A single electric strand 15 to 18-inches high will work as an interim measure to control goats in temporary situations. After a short time they learn to manage it and even, in some cases that I have seen, find that scraping underneath the hot wire gives them an obviously pleasurable "kick." For an electric fence to be fully effective, the goats should be trained to respect it. The best treatment is to make them touch it with their noses. It seems they can quite easily tolerate the current on the main parts of their bodies but not on their noses. Pressing the animal's nose onto the fence (which can often give a kick second hand) or putting some food on the far side, so that the goats touch the live wire trying to reach it, can be the solution. My personal preference is for well built, permanent wire fences with a hot wire as an extra restraint or back up if needed.

Hedges

Hedges would have to be thick and high to contain goats and ideally should have wire fences built in front of them to stop too much damage.

Kids

Fences around kid runs must be good enough to teach them that they can only get to the other side when released from the run — this should train them for life — a very important lesson.

Kids in a secure yard.

Housing and Shelter

All goats need shelter from the weather, either rain or sun. Trees or even rock formations can provide shelter, but some roofed buildings will probably be necessary as well.

The sheds should be made facing the direction least prone to high winds, this varies in different countries and districts — rely on local knowledge. The sheds must be large enough to accommodate all the goats in a paddock. Several sheds are really better than one as often the stronger goats will make sure those low in the pecking order cannot enter their shelters. The horn width of fiber goats must be taken into consideration. Boers' small horns do not seem to be a problem; they are far quieter than fiber goats. All sheds should be high enough for someone to stand upright and a means of closing them off so the goats can be handled inside is helpful. For one or two goats a small, portable, "A" frame or old tanks can be used as shelter, but again, care must be taken that one goat does not stop the others from entering. In Australian country areas where old tanks are easily obtainable, very good shelters can be made by halving them and setting them on frames. Unfortunately these days tanks are made of lighter steel than they were 40 years ago and they might not be as long lasting as they were in the old days.

Dairy

Milkers should be provided with lounging sheds or barns containing racks for feeding hay ad lib in wet weather and winter. They may not always use them, but they certainly will not remain productive if shelter is not available at all times. See Chapter 13 for shed plans.

Bucks

For dairy bucks, separate yards or small paddocks are necessary to stop fighting in the breeding season. These should be divided by a tree plantation or passage and be a minimum of 50 feet square. Each run will need its own house — like the tank shed shown to follow. These runs need to be rested, top dressed with dolomite or lime and eaten off, if necessary, at three or four-month intervals. A horse will clean out after goats and is a useful management resource. They do not share the same interior parasites (and if fed their copper correctly, neither goats nor horses

Raised tank house

"A" frame shelter

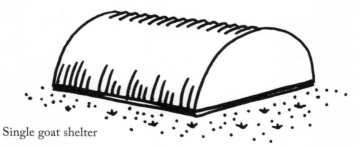

Single goat shelter

Types of goat shelters.

will have them anyway). Several bucks can be run in a group out of the breeding season. Take care that playing does not degenerate into bullying — either the young bucks by the seniors, or, more seriously, an older buck by a strong young one. Should this happen the older buck may pine away and die. This has caused the death of several valuable bucks in dairy studs.

Floors in Houses

Concrete is really the only answer for flooring in sheds regardless of size and building materials. Dairy does have an astonishing output of urine, once reckoned at eight gallons a day — a slight exaggeration I feel. But when the grass is green, goat sheds can, overnight, be reduced to a condition that has to be seen to be believed. I have tried other materials for flooring but I always come back to concrete because it is the only floor which can be totally cleaned. Deep littering is *not* a success with goats, no matter how dry the area or how well set up because it becomes too wet and hot and finally causes problems.

Slatted wooden floors which are high enough to be cleaned out underneath are also good, but expensive. Care must be taken to see the animals on them are not subjected to drafts. Goats like other animals need shelter, it does not have to be very grand, shelter from wind and rain are the main requisites.

Feedlots

The feedlot method has been used for goats worldwide, but I do not recommend it. Laurel Acres in California ran about 2,000 head and managed to last longer than most — apparently the system does not really suit goats. A Canadian goat keeper agreed with me that the lack of exercise was probably the crucial factor in the breakdown of this type of system. On the farm where she worked, sickness and vet bills had reached disastrous levels. After spending a week with my milking herd, which often roamed two or three miles a day, she felt their good health was partly due to this exercise.

The barns housing commercial herds in the United Kingdom seem to work reasonably well, but no one expects the goats to live confined for too many years. The average age seems to be up to six years, then they are sold off and replaced. Thus each doe spends about four years in the sheds. Ideally goats, like all species, must have exercise if they are to remain healthy. The late Frank Thebridge, who bred many high-producing goats in New South Wales in the early days in Australia told me the same thing. He particularly emphasized the importance of exercise for in-kid does.

Meat and Fiber Goat Requirements

These goats need shelter from wind, sun and rain. Fleece goats, both Cashmere and Angora, need better fences and handling yards. Boers, on the other hand, are much more placid and easy to manage.

Choosing a Property

This is more often a matter of finance than choice. But assuming money is not an object, choose a farm with plenty of established trees. If possible, also look for areas of young trees which can be fenced off until they are large enough for the goats not to wreck the foliage. As mentioned earlier, dry land farms are better than wet, swampy ones. Marginal land that is wild and of little use for commercial agriculture will often suit goats very well. They will, however, tame the land in a few years, when it should become part of the regular farm operations. Coastal belts with high humidity and rainfall are *not* good for goats, or other types of stock either.

If there are established fences, they will need to be made goat-proof, or, which is what I do, they can be managed so that the goats have the run of the farm. The larger stock are confined to their own paddocks, this way only the boundary fences will have to be goat-proof. Goats do better when choosing the grazing that suits them, according to the weather and time of year. This can be done by making the rails or top wires high enough for the goats to walk underneath.

Established sheds are highly desirable on any property, placed so the goats can use one or some of them as shelters when they need it. For milkers, the plant and sheds for drafting, milking and feeding will be needed. For fiber or meat goats, shelters, drafting yards and, for the former, a shearing complex will be necessary. Meat goats will need drafting yards and loading facilities and would appreciate having shelter in some form as well.

Goats are complimentary to sheep because they find many plants quite palatable that the sheep will not touch. Cashmeres and meat goats will be fine, but it is important to remember that Angoras must be supervised at kidding time (see Chapter 12). Goats and cattle together need good management unless the farm is fairly large and run on sound organic lines. Cattle in small areas tend to pug up the ground too much and it should be remembered that goats, sheep and deer share the same internal parasites. Years ago goats were often run with cattle as they were reckoned to stop contagious abortion. In fact, the abortion was not contagious, it was caused by a weed, so several cases would occur at once. The goats ate the weed without any ill effects — so no abortion.

As mentioned earlier, horses and goats seem to get on well; the goats will often eat what the horses leave in the way of weeds and different grasses. If you are raising them together, there should be plenty of room and care must be taken to see the horses do not chase the goats.

Tethering for Small Numbers

Tethering *can* be carried out quite humanely, but very rarely is, because constant supervision is necessary. Tethered goats must have access to water and shelter and they should get fresh, untrampled grass two or three times a day. Goats prefer not to eat grass that has had a chain dragged over it or been trodden underfoot.

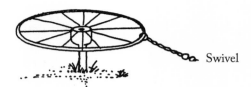
Wheel tether.

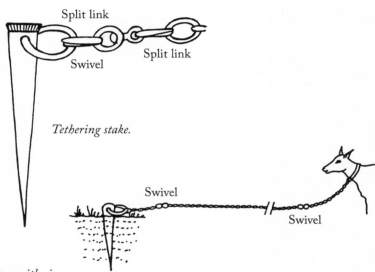
Tethering stake.

Steel peg with ring.

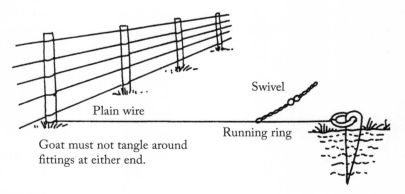
Goat must not tangle around fittings at either end.

Running tether.

Tethered goats are at the mercy of marauding dogs, so it is often better that they are not dehorned in order that they can protect themselves. Tethers can be of the running wire type (see illustration on the facing page). A stake, or a wheel put on a stake, is quite good as it will rotate allowing the goat ease of movement and it can be moved by rolling. Whatever the type, it is very important that there should be a swivel at each end of the chain, otherwise it may get twisted and possibly strangle the goat. When tethering goats make sure there are no obstacles around which the chain can become entangled — trees, loose branches even, can cause trouble.

Yarding

This is probably preferable to tethering, as the goat can move around easily and safely without risk of getting chased or caught up. However, as with feedlot systems, some way of exercising the goat(s) must be found — it is just as easy and often more entertaining to take the goat for a walk as the dog (or both).

The yards will have to be provided with racks for hay and green fodder as well as shelter, troughs for concentrates and a permanent water supply (not fluoridated).

Chapter 3
Different Types of Husbandry

Free Range Commercial

In this system the does are brought in to feed and milk twice daily and left out to graze at other times. Worldwide it is possibly the most usual type of goat farming. The goats are provided hay ad lib and they are fed concentrates either in their lounging sheds or at milking — the rest of their feed is gained from the paddocks. These can be rotated or the goats can be given free range over the farm where they will choose the areas they prefer each day. As described in the previous chapter, lounging sheds or shelter with hay provided must be accessible at all times. Milk production will suffer if the goats are under cover without food in rain or excessive heat.

Indoor Commercial

Countries with snow and icebound winters have to confine their goats — years of this kind of husbandry have made the farmers very adept at it. In the United Kingdom, United States and Europe, some commercial units house the goats all year round; this is different from feedlots where the goats are confined in yards with hay, feed and

Free range goats roam our property in Australia (above and below).

water along the sides under cover. Indoor commercial farms provide hay ad lib and have troughs outside the fence for concentrates. Some farms feed while milking, but most do not as the slower eaters are disadvantaged.

The two farms I saw in the United Kingdom that practiced this type of goat farming had large, airy sheds with slatted sides that could be closed up when cold. Both housed about 200 animals, divided into four lots for the purposes of milking and cleaning out. These farms had the advantage of a large, highly trained staff to run them. The one in Sussex had the processing facilities for the milk on the same premises and a staff of 13, counting the manager and his wife who also worked. All the fodder was grown on the farm and handfed to the goats and only the goatlings were allowed controlled grazing. On the other farm at Corfe, in Somerset, the interior arrangements for the does were similar — again the milk was handled and processed on the farm. The staff consisted of a very efficient couple who ran the goat operation and who called in labor from the main farm when necessary.

Similar arrangements would be difficult in Australia due to the cost of labor, the award farm wage in the United Kingdom at that time was the equivalent of $250 in Australian money a week ($158 American), less with a house provided. Both farms used employed labor for the goats, the owner looked after the main farming enterprise.

Indoor goat operation.

Keeping sheds clean and sweet while allowing the animals comfortable bedding is always a problem. The farm I saw in Sussex managed it well. The shed, which was about 200 x 100 feet, was divided into four with a cross-shaped passage in the center, each section holding 50 goats. One section was drafted out at a time to the holding yard and, once or twice a week, a tractor and blade removed all the bedding from the concrete floors and put down new bedding while the goats were being milked.

One of the advantages of these systems was an unending supply of high quality manure for top dressing the paddocks. This was either composted and spread in late winter or put straight out, working across the farm. This practice maintains the fertility of the farm and any goats that were allowed free grazing (like the goatlings) avoided the freshly spread areas for about three months. This allows time for the dung beetles and worms to remove the contaminating feces below the surface where they are "processed" by the soil bacteria. The lime levels should be monitored regularly on farms treated this way, and adjustments made when necessary.

In some complexes the goats are housed on extruded metal when under cover, this allows the manure to drop through onto a concrete floor below and from there it is hosed out. Goats do not appear to like lying on these floors, nor can they be fed on them as the floor would become clogged up. These sheds seem to be permanently damp in the winter months and indeed pneumonia had been a problem on one farm in Australia that used this method. Feeding was done in feedlot-type yards outside, which works reasonably well in mild weather, but causes trouble in the cold and wet.

One fact which makes any commercial operation in Australia totally different from those in Europe is the population numbers, and vast distances between population centers. Both the European operations collected any extra milk they needed within a 40-50 mile radius of the farm and marketed all their products within the same area.

The first farm used hormones to control breeding so that they had a year-round milk supply. The second used lights as described below very successfully.

The disadvantage of permanently shedded goats is the limit on their useful milking life, this usually only lasts from the first lactation until six years at the longest — generally less. On free range or limited access to the farm conditions, the goats live longer, milk longer and the cost of rearing replacements is reduced.

Goat Farming in France and Spain

Goat farming is big business now in Europe. An Australian goat keeper who worked in France reported that in the Loire Valley alone there were about 40 goat farms. These either ran the goats fully housed or with controlled grazing. Cheeses were their main sales, either in local markets or sent to be sold in Paris, the population of which is around 15,000,000 so the market is virtually unlimited. This highlights the problems that are inherent in Australia, as in the United Kingdom, the French rarely travel over 50 miles (generally much less) to market their products. While in Australia, distances between selling points is much greater.

In Spain the ultimate goat dairy is now operational. The does all have computerized sensors on their collars that record the amount of milk they produce. Their feed is allocated on this amount and they may only put their heads into the manger that corresponds with their computer read-out. This sounds complicated, but apparently it works very well and the method is widely used in cow dairies.

Alternative Methods of Controlling Breeding

There are various methods of persuading goats to breed out of season. These include the use of hormones to synchronize estrus (usually for artificial insemination) and to vary breeding cycles. Lights simulate early summer and also an early autumn to persuade the goats to start ovulating two or three months sooner than usual. This method has the same effect without the inherent disadvantages of hormones.

Hormones are often used once or twice by many goat keepers, but seem rarely to be successful as a long-term husbandry resource. Laurel Acres in California abandoned what seemed to be a successful artificial insemination and controlled breeding program for no very clear reason, obviously it was not working as well as was expected. Artificial breeding can also be expensive and the success rate seems to vary considerably in goats.

Lights, on the other hand, merely "kid" the goat's system into ovulation earlier by simulating an early spring or summer. A large quartz halogen globe in each corner of the big shed (approximately 66 by 98 feet) housing the milking does is used, they produce enough light (in English conditions, which are rather dark compared to Australia) to simulate an English summer. Whether this method would work everywhere is questionable and it does mean that the goats have to be permanently housed, which, for reasons given above may not really be desirable.

Goats that have had hormones of any kind used on them — either to synchronize estrus or for ovum or embryo transplants — will be found to have vitamin A deficiencies and sometimes these deficiencies occur at quite serious levels. Either the hormones used during these artificial processes destroy vitamin A in the liver where it is stored long-term, or they interfere with the actual synthesis of the vitamin — sometimes permanently. This is the reason why goats that have been on artificial breeding programs, particularly if they are on dry country where building up their reserves of vitamin A would be difficult, often have problems returning to normal breeding patterns. A lack of vitamin A is the biggest cause of failure to conceive in all species.

As long as this fact is recognized and the goats (whatever the type) are supplemented regularly with some form of vitamin A, artificial breeding should not have too many side effects. Hormones also interfere with the assimilation of calcium and magnesium.

Artificial insemination, on the other hand, should work well, but for some reason there seem to be problems with goats. Laurel Acres, mentioned previously, gave it up as did a concern in Canada under the auspices of the University of Guelph. The operation abandoned the method after struggling with it for a few years, no reasons surfaced. In Australia artificial insemination (A.I.) has had mixed success, possibly due to the fact that times of insemination have not been ideal. In C. Gall's work, *Goats*, one paper puts the best time for A.I. as twelve hours after the start of estrus, when there is a 75 percent chance of success. Every five hours later decreases the chances by 25 percent. Obviously A.I. is of great value, sires that would be unobtainable for a breeder can be used thus widening the gene pool or for selecting certain lines. If using A.I. heavily care should be taken that the gene pool spread is maintained. In Canada, when A.I. was first used for cattle, Hereford breeders suddenly realized that every cow in the country was related to one of nine bulls used for the scheme. The animals were in danger of becoming too inbred.

Extended Lactations

This method of extended lactations works well and is in use in several commercial dairies here. I ran a small commercial unit for many years using extended lactations with great success. It is also the preferred arrangement for small goat keepers who have two does, kidding them alternate years and thus ensuring a year-round supply of milk. Not *all* goats will milk through — all British Alpines will, if properly fed and looked after, and so will some of the other three breeds. The first cross of the Alpine with any other breed carries this characteristic. Anglo Nubians used to be famous for their extended lactations 100 years ago, but now it is sadly not the case. The really good ones still have long lactations, but they are in the minority. As the British Alpine is a type within the French Alpine breed, it is probable that they carry the long lactation gene as well.

The advantage of this method is that the goats are rarely dry — approximately one or two months in each 24. They are kidded at 14 months then milked for two years, and so the cycle proceeds. Half the herd is mated each year. The does that are running through will drop their milk a little in the winter as all goats habitually do, but when the other half of the herd is kidding, their milk will rise to something very near the fresh kidders supply. Little arithmetic is needed to figure out that over a number of years this method provides more milk than when the does are kidded fresh each year. The strain on the does is less and they will often milk quite steadily till nine or 10 years and beyond. Kids born after a two-year gap also appear stronger and more viable than the produce of yearly breeding.

If long-lactating goats have been bred every year for some reason — usually a desire for more high-priced kids to sell — they will take a couple of years to settle down to running through the winter.

Mountain Husbandry

Mountain farming is practiced in Scandinavia, Switzerland, Germany, Austria and other mountainous countries. The goats are depastured up the mountains in the summer, taking their milking plant with them. Usually the mountain herbage is of such high quality that little extra feed has to be given while they are on it. In the winter they are brought down to the valleys and housed in barns rather similar in style to the places I saw in England.

Occasionally family goats, like family cows, are housed up in the mountain in chalets for the winter. These are filled with good mountain hay during the summer and the animals eat their way through it during the winter months. This way their owners keep a milk supply handy, and have a supply of dung for the chalet garden. The goats and humans often share the same building in these colder regions — although not the same part of it.

Herding

This is the oldest type of goat husbandry in the world and is practiced in many places where labor is cheaper than in developed countries. One (or more) herders drive or lead the goats to the grazing areas as soon as they have been milked in the morning. In the evening they return to their enclosures, are milked and stay in until morning — safe from marauders who would like a goat supper. In Malta, where grazing is sparse, the goat herder carries the animals' midday meal of concentrates out to the pasture. This system, though largely impracticable in many countries, is probably an ideal one. Goats relate very much to people and regular routines.

I have tried trucking my goat herd to good browsing areas, especially in drought. It did not work, they flatly refused to leave me for more than few yards and would not eat — the hypothetical lions were too dangerous. I finally found it easier to cut the bush and bring it back to the goats.

Goats on Small Holdings

A large part of the world's goats are owned and farmed in small units. Four or five goats can supply the needs of a family for milk, meat and skins. The animals relate to their owners on a very personal basis. They are usually kept in a small paddock with a lounging shed and are almost entirely hand-fed. Sheds should ideally be high enough for the owner to stand upright in — it is easier than tending goats on hands and knees. In England at least, these goats are taken for walks daily to browse hedgerows and gain much of their feed this way.

Hobby Goats

Many people have goats as a source of milk and to enjoy showing them. These goats usually have the best lives of all and are often outrageously overfed, especially when young. This is because goats, unlike any other livestock, are shown as kids and goatlings as well as when they are milking adults.

Nevertheless this is a rewarding occupation that can involve the whole family, especially when children are young. My children really used to enjoy taking a car load (14 once) of goats to shows and all the preparation that went with it. So the family milk producers provided us with amusement at weekends.

The ideal way to keep these "hobby" goats is to give each one its own stall for feeding and resting, letting them graze during the day with a lounging shed in the paddock.

Stud Breeders

There is definitely a place for the specialist breeder who goes to a great deal of trouble to fix good milking lines. In the United Kingdom and Europe these breeders supply commercial concerns with male and female stock to improve their herds. Commercial farms breed a certain number of their own replacements and sell young stock and old goats who have ceased to be viable (at about six years). However, the time always comes when new, proven bloodlines are needed; this is where the specialist breeder comes in. They have time to show animals, thus ensuring the replacements they sell are sound in conformation as well as being high producers. In the United Kingdom the emphasis is on the milking ability of the animal, and at most shows goats have to gain a "Q Star" (see page 332 for definition) to be allowed in the inspection classes. A fact often forgotten is that dairy goats are a milking animal.

In Australia during the post-war years there were apparently a great number of high-producing goats. Unfortunately their physique did not always match their production. I remember seeing several does with grotesquely large udders — they could hardly move — very undesirable. One of the breeders of that era told me that the most frequent troubles were udder injuries — does stood up after resting with one foot on the udder with ghastly results.

Fiber Goats

Fiber goat farms vary from the highly specialized breeding establishments, who partially hand feed their animals and produce high-class pedigree stock for herd improvement, to the broadacre farmers who still use good bloodlines, but treat their goats as sheep farmers do. Both methods have their place. In the fiber breeds, the show and the range goats tend to differ in the amount of facial cover, as they do in South Africa. This appears to me to be a double standard, which does, however, seem to work in practice.

There are also a significant number of home spinners who keep fiber goats in numbers ranging from two or three to 30 and up. Some goats are colored, and the artifacts made from both mohair and cashmere are generally of high quality.

Meat Goats

Until about six years ago this meant ferals and fiber goat culls. It was a chancy market, and never began to satisfy the huge demand for goat meat (*chevon*). Australia is surrounded by nations whose first choice of meat is goat, either kids (*capretto*) or full-grown animals, yet little had been done up until that time to supply a large and continuing demand. When I was Victorian State Secretary about 15 years ago, there were requests coming in from Japan quite frequently; particularly for bulk supplies of goat meat. No one was the least interested in production for this market. In the United States, most of the meat markets will be on shore and distribution will be easier.

The Middle Eastern countries also prefer goat. In the early 1990s the first Boer goats arrived — called purpose-bred meat goats — and these differ quite markedly from the usual idea of goats as seen in this country since the time of settlement. All these were, unless well handled when young, inclined to be flighty and "capricious." Then the Boers suddenly offered serious meat breeders an alternative, which even in the first cross on dairy (Anglo Nubian) or fiber goats was placid and a good doer if the land was properly balanced. Even kids reared on their dams appear docile and placid and the mainstream farmers found themselves with another source of income which fits into the general livestock picture without specialist fencing and facilities.

Nowadays Boer goats, either black and white or in some cases chestnut, are being farmed for chevon or capretto. The numbers are still being built up worldwide and either Nubians (which must share a common gene with this breed), Cashmeres or Angoras are being used for upgrading and, of course, other breeds for ovum transplants.

Chapter 4
Acquiring Stock

This can be quite a difficult undertaking whether the stock is needed for dairy, meat or fleece farming. Before the advent of CAE (Caprine Arthritis Encephalitis) it was not such a problem, but now buyers *must* ensure that they acquire CAE-free stock. This disease, which is described in Chapter 11, is an incurable auto-immune disease which can, like other illnesses of a similar nature, be "carried" even though the carrier may show no signs of the malaise. Therefore, the appearance of a herd or a goat is *not* a reliable guide. If the breeder/farmer has carried out a reasonable program of eradication and/or testing for CAE, and the documentary evidence shows the herd or goat is clear, the buyer can go ahead.

These days there is another concern — Johne's Disease or *Mycobacterium avium subsp. paratuberculosis*. It is not as readily transmitted as is generally believed. It can be cured if the animal is valuable enough to warrant the trouble and it can certainly be prevented by seeing that the paddocks are in top order and the goats are fed correctly — receiving all their minerals, especially copper.

Do not be tempted by doubtful stock just because it is cheap and looks alright, the CAE virus can lie dormant although Johne's-affected animals usually look below par. In any case, buying trouble is *not* an option, have tests done by your vets to confirm that the stock is sound. With CAE, as long as the goat stays on the farm where it was bred and is not subjected to the stress of moving, it may show no signs at all. However, as soon as it reaches a strange environment on another farm, the stress caused by the change will almost certainly cause CAE to flare up. The only possible reason for buying, or preferably acquiring free, a known reactor to the disease is if the goat is a rare and valuable bloodline. In this case it may be worth trying to bring the doe through one kidding at least — the kids may be "caught" and hopefully (following the suggestions in Chapter 11) the doe may survive for another kidding.

Bucks do not usually transmit either disease, at first this was not realized with CAE and many good bucks were put down. Positives can be used and a strict testing regime should be used with the resulting kids. One word of warning — the kids of female positives are never quite as robust as those of clear, healthy goats. It seems to take two generations to get really tough progeny.

Acclimatization

This is another factor that must be considered when buying stock. Goats from latitudes nearer the equator will adhere to their early breeding times for their first year in the north, and after that first year they will fall into line with the local breeding time. Quite often goats from another district or state do not "do" very well, but their descendants will be fully acclimatized and present no problems whatsoever. The new owners of the Texan Angoras seem to have few problems, but they are the third generation born in this part of the world and are fully acclimatized. But even these fully acclimatized goats show a preference for healthy pastures with good magnesium levels, and this should always be remembered. Coming from Texas, these animals are used to good magnesium levels in the soil, something which is rare in Australia.

Price

This is governed, as always, by supply and demand. At the time of starting the first edition of this book (1989), good milking stock fetched anything from $100 to $1,000 Australian. Two years later the demand was almost non-existent for dairy as well as fiber goats, but by 1998 signs of recovery were in sight. The release of the Texan and South African fiber goats early in the 1990s, along with some very hard work by members of the Angora Mohair Breeders Association, have revitalized an industry hard hit by the recession. Cashmeres are, as always, holding their own.

But the usual scenario in dairy goats of commercial setups starting and then stopping equally quickly still seems to be with us. A few stalwarts hang in there and the cheese producers are possibly the backbone of the industry. During the last 30 years in goats, large swings in prices have occurred with almost monotonous regularity. Boer goats are now providing the impetus and should fetch good prices as an adjunct to the meat industry — we need the exports to countries that eat goat by choice.

Show goats fetch prices that are completely unrealistic as far as commercial set-ups are concerned. A very good milk buck should make his price, but only as the source of a new and useful bloodline. At the time of writing this edition, Australian show goats are changing hands for $500 up to $1,000 USD.

General

As above, when considering the purchase of any goat, have your vet take blood tests for CAE and Johne's disease; check a prospective purchase for lice, bad feet and general good health. Pull the lower eyelid down to see if the goats are anemic and ask the seller what drenches and medication may have been used on them. Many of these health problems will respond to good husbandry and are not insurmountable, but the price should be adjusted accordingly if the animals are not 100 percent healthy.

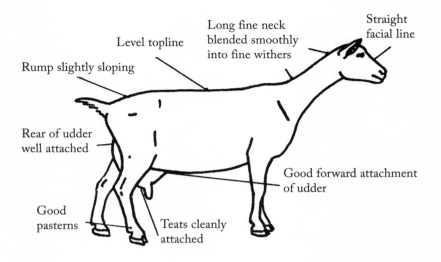

A good type of dairy goat.

Age

Check the teeth for age. At one-year old there will be eight even teeth, at two years there will be two big ones in the middle and three of the small ones on each side. If a goat is very minerally deficient these "two tooth," which is a description of the age, may not have come through. At three years there will be four big teeth in the middle and two small ones on each side (known as "four tooth") and at four years and after there will be eight full size teeth, known as a "full mouth." From then on the size of the teeth may vary; for instance, goats on hard grazing or who have to eat lichen off the rocks as many do in drought, have their teeth worn nearly flat with the jaw. As long as they are healthy and properly fed, these teeth do not get loose as is sometimes seen in aged and improperly fed sheep.

Conformation

A well made goat is just that, whether it be for fleece, meat or milk. A leg is needed at each corner, to quote a well-known judge. Those legs need to be strong and

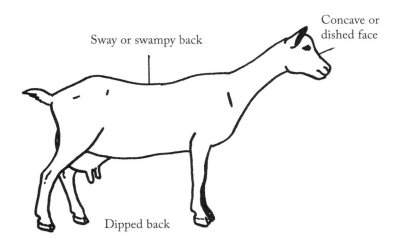

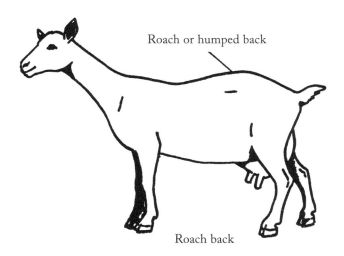

Back problems in goats.

straight, no matter how good the milker or how fine the fleece; if the animal cannot move around soundly, it will not be worth buying. That said, it should be pointed out that a fair proportion of cow hocks, bad pasterns and crooked front legs are, regrettably, environmental not genetic; there are still goat keepers out there who do not know how to feed an animal to ensure sound and healthy bones.

A buyer would have to decide which cause was the real one if confronted with a weak-hocked animal. A good "milky" type of doe has a "look" about her, rather like a good Jersey cow; she is not coarse, has a fine neck, good skin and an alert expression.

A strong straight back is essential. A weak back, often looking as though another pair of legs is needed in the middle, will not carry successions of kids or a large udder for very many lactations. Unfortunately this particular fault appears to be on the increase. A roach or upward rounded back is not pretty, but it is reasonably functional, although it may be penalized in the show ring.

The chest should be wide. Narrow chests tend to creep into goats at certain times and breeders should always endeavor to breed a good broad-chested animal whose front

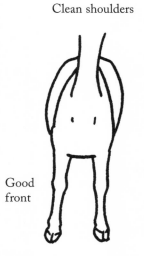

Chests: Good (left) and bad (right).

legs are neatly blended into the shoulders, not stuck out like wings. An adult goat should have at least four to six inches between the front legs, any less than this denotes diminished lung space and an animal prone to problems in that area. Should a very narrow-chested goat become infected with lungworm, it could easily die as there would not be enough lung area left for it to gain the oxygen it needs. Goats have small lungs; they are, after all, a grazing animal and not one that lives by speed as a horse does.

Wry Tail

This is not good and many breeders associate it with a crooked spine, although I have not found this to be true. Wry tails are occasionally found in Nubians and are not considered a fault in the breed.

Wry tail.

Mouth and Head

All animals' mouths should be checked; the teeth should meet the upper pad squarely (goats, like sheep, only have front teeth on the bottom jaw) and should not be over or under hung. A protruding bottom jaw is occasionally found in the Nubian breed. It was quite functional when they were in their native land and had to browse above their heads, but here it does not help when grazing grass and is not often seen now.

Occasionally jaw abnormalities can arise as a result of malnutrition. Many years ago I sold a normal-mouthed kid to a district about 60 miles north of my home; when later

I went to live in the area, the kid I had sold was brought to a buck for service with a badly deformed bottom jaw. I did not actually believe it was the same goat until I had checked the tattoo. I agreed to let the buck cover it on the condition that it and the kids were to be shot if the defect showed up in the kids. A year later the owner came again with a normal-mouthed doe for service; I expressed my relief and asked what happened to the other one. What follows was the answer. "Oh well you are always rabbiting on about dolomite and I had never bothered with it, so I got on and fed it, the mouth returned to normal in six weeks."

That said, it is 99 percent certain that heredity is the reason a kid is born with a deformed jaw; it does occur far too often. The mating that produced the kid should not be repeated and the kid should be destroyed. If the condition was not hereditary, good feeding would cure it. This has happened in horses born with bone defects caused by malnutrition *in utero*.

Worn teeth in older animals, provided the teeth are not loose or pointing in different directions, are alright. If goats have lived through droughts or on hard country their teeth will often be worn flat to the jaw, but provided the wear is even and the teeth are not loose, it does not stop them from grazing. If they have been fed their minerals correctly,

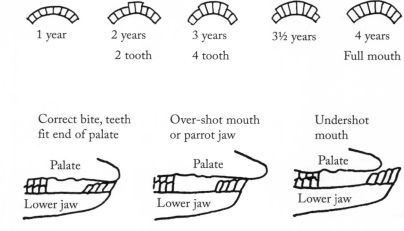

Teeth and jaws in goats.

goats (or any other species) should not ever have loose teeth.

Wry Face

There has been much controversy on the exact classification of a wry face. The spine down the center of the face should be straight and the eyes level.

Wry face.

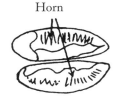

Horn

Neglected foot showing horn bent under foot

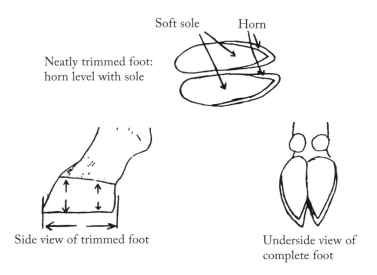

Soft sole   Horn

Neatly trimmed foot: horn level with sole

Side view of trimmed foot

Underside view of complete foot

Feet.

Acquiring Stock   53

Feet

The feet, provided they have been kept trimmed, should be straight and even and without a wide gap between the toes. The state of the feet is more often indicative of the husbandry than inheritance. If the goats have lived in an area where there are plenty of rocks, their feet will have been kept in good order without having to resort to the trimming shears every six weeks. Good, straight feet can be hereditary, so when buying a kid, examine its mother's feet.

Dairy Characteristics

Here we diverge between the dairy, meat and fleece types. Dairy goats need a capacious rumen, so that seen from the side the body is a wedge shape from front to back and seen from above the body is a smooth double wedge from the shoulders to the rump. The muzzle should be broad — a pinched muzzle often denotes a poor doer. Fleece goats are more active and light, so are inclined to be a series of rectangles rather than wedge shaped, and Boers are similarly built, only on much more generous lines.

Good face   Pinched muzzle

Muzzles.

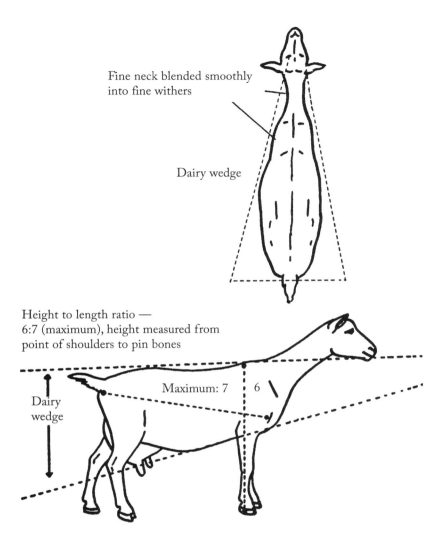

Dairy goat ideal wedge shape.

Fat Goatlings

Some conformation faults arise from animals being allowed to become too fat as goatlings, often for reasons such as showing. A goat's forelegs are not attached to the frame, but are held in position by muscles and ligaments. If the animal becomes obese, fat will be laid down between the shoulder blades and the body. Worse still, it will also be deposited in the udder area producing a fleshy, poorly producing vessel. It is extremely difficult to remove these fat deposits and they should not have occurred in the first place.

Rump

A reasonably flat rump is desirable. A short, steep rump does not, as we were often told years ago, cause kidding problems — but it does generally denote a flat, ill-attached udder underneath it.

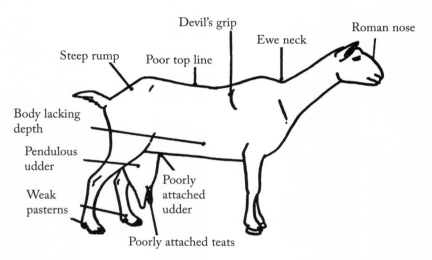

Poor conformation.

Back Legs

These should be straight below the hocks, whether viewed from the back or side (as in the illustration on the next page). Sickle or too straight hocks are a weakness best avoided, both tend to break down with age. Nubians should have straighter back legs, showing almost no bend at the hock as in the "Swiss" breeds and this does not appear to be a weakness in this breed. Meat and fiber goats have to carry heavy fleeces or body weight and must have straight sound back legs.

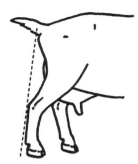

Correct rear angulation for "Swiss" goat

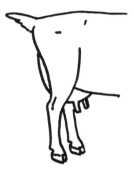

Too straight and vertical

Sickle hocked

Back legs.

Udder

In a milk breed the udder should start well forward of the stifle, be capacious and run smoothly back with well-attached sides between the back legs to a good, wide, rear attachment. Both sides of the udder should be even, although there are few goats that do not show a slight (often invisible) bias to one side or another. There are countless variations of this desirable shape, some worse than others (28 years of milking commercially by hand taught me a lot about udders). When milked out the udder should feel soft and empty. The udders of very large producers will naturally have some tissue apparent, otherwise their udders would not survive one lactation, let alone two or three. But a vessel that, when milked out, still looks and feels almost the same as it did before milking denotes a poor producer.

When breeding for good udders, my criteria are that the udder will stand up to many years of milking without altering or breaking down and that it should have a good flow, making it easy to hand milk (and therefore machine milk also).

Udders of fiber and meat goats should be free of deformities and have teats that the young kids can easily suck. They will not feed their kids for more than three months as a rule but, if they have a reasonable supply of milk, the kid will grow on better than if they are poor producers. In the fiber breeds, extra teats are frowned upon in the show ring, as are either supernumeraries, sprig or double teats. There is actually a sound reason for this, see the next page.

Teats

The teats should be straight, easy to hold or attach to milking cups and point down or slightly forward. They should be at the bottom of the udder, not half way up the sides, which makes milking either by hand or machine difficult. The teats should allow the milk to flow freely once milking has begun — thin, tight teats can be hard work for hand or machine.

Side view

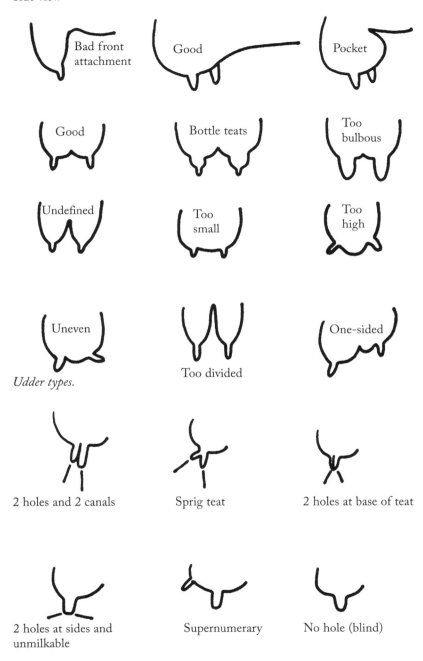

Udder types.

Teat deformities.

Obviously the previous remarks apply to a lactating goat, but even in a kid or goatling careful examination of the teat area can give one an idea of the udder's future appearance. If the teats are very small or close together, it may mean an unproductive, narrow udder. Each teat should have *one* hole, and there should only be *two* teats. Deformed teats are a very serious defect and the animal should not be considered for purchase at all. This deformity is usually penalized in the show ring.

Supernumerary teats are not quite such a matter for concern provided they are not too near the main teats. They are a disqualification in breed societies and therefore in the show ring. But provided the extra teats are not removed until the doe is actually milking, the goat can be a perfectly useful house or commercial milker. The reason for not removing supernumeraries at birth is that they can occasionally have a milk canal, in which case one would be left with a leaking hole when the doe came into milk. Worse still, the removed teat could have denoted another quarter in the udder, which would need to be milked out.

I have seen this phenomenon two or three times in several breeds. Evidently there was a strain that produced double teats way back and these goats were not destroyed when the deformity was discovered (as they should have been). The goats that I saw had normal looking udder development until they started lactating when it became painfully apparent after milking that there was a third quarter still full for which there was no teat.

Another phenomenon that I have seen twice, and other breeders have also seen it, is the kid that is born with normal teats and later develops another — both the cases I have seen were minutely examined at birth by independent inspectors and/or breeders. Five or six weeks after the initial examination another teat has grown down beside one of the normal ones — with another hole. If the kid is a breed in which there have been teat problems, it would be wise to make quite sure that it is two or three months old before selling it. This latter phenomenon arose in Australia in a breed with a woefully limited genetic spread.

Scrota and Teats in Bucks

A buck's scrotum and teat attachment will give a reasonable idea of the kind of teats and udder shape that he will throw. Uneven scrota should be avoided; bucks with this defect have a nasty habit of throwing one sided or very uneven udders. Each side of the scrotum should have a descended testicle in it, a buck with a one-sided scrotum (cryptorchism) should be destroyed at birth. This applies to all goats, meat, fiber or dairy. One of the main reasons for this is that undescended testicles can sometimes be in peculiar places and cause pain, they can also make wethering (castrating) difficult to impossible.

Flat scrota, instead of being well rounded (and nothing to do with the ambient temperature which will make scrota drop on hot days), tend to produce flat-fronted udders.

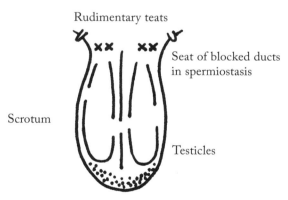

Buck's scrotum.

Spermiostasis

This is an hereditary defect in some polled lines of dairy goats, especially if the sire is pure for poll. It can be detected, except in very mild cases, by palpation of the scrotum. This should feel firm and even without any hardened or necrosed areas along the top. If in doubt, have a veterinary surgeon check the animal before buying. A buck with this complaint could possibly get one or two does in kid, or none at all depending on the severity of the condition. In spermiostasis, one, two or all of the tubules along the top of the scrotum are closed, so the sperm cannot be delivered. In the case of only one or two being closed, there may be a few successful matings, but the blocked area will gradually atrophy and cause the remaining tubules to do the same, rendering the buck infertile and giving the animal great pain.

There is a connection between cystic ovaries in does and spermiostasis in males. Lines suspected of carrying either defect (always polled) should be avoided except by very experienced breeders. My best line came from a doe whose father only achieved two successful matings before he went infertile with spermiostasis and she was always bred to disbudded bucks. She founded a great line of milkers — the polled lines generally run to higher milk.

During the 1930s, Saanens in Germany (where only polled goats were permitted to be kept) became a breed which was infertile in large numbers and when the cause was discovered, the law was changed so that only horned kids could be kept. Of course a mixture of the two is desirable, especially as polled lines often carry the highest milk production.

General

Many of the above observations apply equally to meat, fleece or dairy goats. Faulty teats in bucks or does are to be avoided, and faults in the scrota of bucks should be treated with caution.

Fleece goats should have an oblong body with good straight legs "at each corner" — the body should feel solid and square to the touch. A doe with good udder conformation and capacity is obviously going to rear better kids than one who is lacking in this respect. Well made feet are of particular importance in all goats, as they will often have to live on rough country and travel far to find their food.

Fleece

Angoras should show good character and style and be free from kemp as far as possible. If it is feasible, a micron count should be taken of any stock that is to be bought for breeding purposes — both bucks and does. The fleece is after all the marketable commodity. It should be remembered that fleeces tend to deteriorate with age and if a buck or doe of advanced age shows good fleece characteristics, it is probably very good indeed. A friend who was upgrading her Angora stud managed to obtain two twelve-year-old does of impeccable Angora lineage. They looked like short Saanens, no sign of fleece anywhere at that age, but she bred up a beautiful herd from them.

Look for goats that have reasonable belly and face cover, but too much of either can be a drawback in rough country. Facial cover is a matter of choice. South African stud Angoras show excessive facial cover, to the point where they would be blind in range situations. The fleece animals on the farms have clean faces, as they obviously must to survive. But it does come perilously close to a double standard. The extra grease found on Texan Angoras makes evaluating their fleeces a little more dif-

Angora goat.

ficult at first. They do however, seem to be totally free from kemp.

Kemp or Medullated Fiber

This is hair as opposed to fleece — it cannot be spun, died or otherwise processed and brings down the price of the clip very materially. It also can be seen occasionally in the finished product which is most undesirable. Kemp can often be detected by studying the top and base of the spine, the backs of the legs and front of the shoulders. Kemp (hair) can be a persistent problem on occasion and very difficult to breed out. Angoras that have been upgraded from very carefully selected ferals — the set of the horns will tell the experienced buyer whether the ferals carry an Angora gene — are usually the most satisfactory. The guard hair, as found in Cashmeres, should be avoided — it is as persistent as kemp, but of course normal in Cashmeres.

The British Saanen types often used initially for Angora upgrading did not do much except leave a legacy of kemp. On the other hand, the pure, old Saanen type which was of Swiss or French origin is smaller, sometimes having a soft, silky fringe, and will often produce quite creditable Angora types in a generation or two. Ideally, of course, the purebred Angoras are best, but initially they were too expensive for many would-be breeders. With the fall in Angora prices in the late 1980s and early 1990s, and the availability of the new imports for crossing into herds these days, no one should consider buying anything except purebred Angoras.

Cashmeres

These should also show good cover, especially on the belly and around the chest and neck, where the fleece should be soft and silky to the touch, *not* hair. Here again, the tendency is for facial hair to be more obvious in the show ring and perhaps this double standard that exists in both fleece breeds is not really to the benefit of the industry.

Cashmeres can be upgraded from ferals as the guard hair is quite acceptable. The nucleus of a good Cashmere herd can be obtained from carefully selected feral goats, indeed this is how the present herd in Australia was mainly built up in the late 1970s and 1980s. This breed was, like the Angoras, imported into Australia in the 19th century and Sir Sidney Kidman owned large flocks. When this great grazing empire broke up, the goats eventually became part of the feral herd which explains why careful selection of ferals using the horn type as a guide can often pay dividends in good type animals — the genes of both breeds are quite often still out there somewhere.

Cashmere and Angora flock.

Horns on Fleece Goats

There have been fleece goat breeders who have dehorned their herds for ease of handling just as dairy people do. This is feasible in areas where predators are not a problem. Generally the practice is not encouraged as the set of the horns and their shape is considered indicative of the pureness of the breed. If horns are allowed to grow on milking-type goats, they generally grow straight back over the body, without a twist or spiral — unlike the upward and outward curl of Angoras and the less exaggerated version found in the Cashmere.

Boer Goats

The same rules for sound conformation and freedom from faults that could cause problems, of course, apply to Boers. From what I can see of the industry now there are enough individuals for a healthy genetic pool.

So many of the new species starting up in Australia have been hampered by too few individuals initially, but the Boers were imported in sufficiently large numbers to be considered a viable breed. The skins as well as the meat of the breed are obviously very useful, even with the hair on as far as the pure breeds are concerned. It may take a generation or two to lose the fleece characteristics for those upgraded from fiber goats; however, those upgrading from Nubians are seeing results sooner, as one would expect.

Red Boers.

Chapter 5
Breeds

There are many breeds of goats, some more pure than others, worldwide. In certain localities the genetic strains are fairly true, in others they have been totally lost. There is also a large majority of indeterminate crossbred animals. The Rare Breeds Trust in the United Kingdom and the American Livestock Breeds Conservancy are carefully gathering individuals of lost breeds and endeavoring — apparently with some success — to breed these once again.

A rare breed of sheep from the Isle of Man, Manx Loghtan.

The domesticated breeds range from the Pygmy and Boer goats in South Africa, to various breeds with longer and longer legs as one goes north in the continent where the main forage grows high off the ground. Both Boer and Pygmy goats are now part of the goat keeping scene worldwide (the latter are not yet in Australia), Pygmies being a favorite pet goat in Western countries and the Boers fulfilling their potential as good meat suppliers. The Sudanese goat is an example of the long-legged goat and is not unlike the French and British Alpine in conformation if the photographs are to believed. Then there are the Red Sokoto goats of Zambia. As you come to the sandy regions of North Africa the long-eared goats are starting to put in an appearance, the ears of the Mubende goats droop, as do those of the Maltese and Damascus goats. Here we are starting to see the ancestor types of our Nubians, the Zaraibi in particular. This breed was crossed with the Jumna Pari of India to make the breed we know today. The reason for the drooping ears is to keep the sand from the ear orifice and also from the eyes, Swiss-type goats and La Manchas with erect or tiny ears would not enjoy a sandstorm.

Boer goats.

Pygmy goats at show in Britain.

Most of the above-mentioned goats serve a triple purpose: milk, meat and hides. They are found in communities where all three are much valued, more than they would be by the westernized nations who tend to place most emphasis on the milk. Fleece-bearing goats are found in the higher, colder climates further north, in what we know as the Middle East.

In the United States and Australia we are concerned with the five milk breeds, two fleece breeds (three if one counts Cashgora), the two meat breeds that comprise our indigenous ferals, the imported Boers and a pet breed, Pygmies, which are to be found in most countries except Australia. The Swiss-type milkers are the Saanen, Toggenburg, British Alpine (similar to the Sundgau type in the French Alpine breed which is not found in Australia), and, of course, French Alpines, which are used for dairying in the United States and worldwide. Then we have the Nubians whose ancestry has been touched on above.

In the early 1980s I had a girl from a big commercial dairy near the University of Guelph in Ontario, Canada staying for a few weeks. She never ceased chuckling about the oddity of taking one of the colors from the French Alpine breed and founding British Alpines as had been done in Britain and Australia — especially as the exercise has been plagued by problems related to an inadequate gene pool (predictably, as it was founded from one goat). She did however admit my herd produced what was wanted.

In Australia there is a large and wide-ranging herd of feral goats, which wreaks considerable havoc by killing vegetation, causing erosion and desertification. Similarities with every breed so far mentioned may occasionally be found in these herds; some look exactly like Spanish Mamber goats, some are marked like Boer goats, and at least five percent are like the Schwarzhals (black neck) type of Europe. Many quite obviously carry the Angora gene and a far larger number are of Cashmere type, these are descendants of the goats already mentioned which arrived in the early days of settlement.

Angoras

A little of the early history of this fleece breed was sketched in Chapter 1. There is little doubt that some of the so-called Cashmere herds mentioned in early shipping lists did, in fact, refer to Angoras. As was seen in Chapter 2, the same misapprehension occurred in the United States in the 19th century. Even today in Australia several quite distinctive types of Angora, dating back to the foundations of the breed, are occasionally to be seen.

White fleeces with the lowest micron count are the most desirable and therefore fetch the best price, but there is also a demand for colored fleeces, especially for home spinners. Kemp, which is hair as opposed to the silky fleece, has been a problem in Australia for the reasons discussed in Chapter 4. Some of the best Angoras I have seen were upgraded from the old Saanen types mentioned previously, and were, in two generations, almost indistinguishable from the pure-bred Angora. Hard though it may be to believe, in the United Kingdom one enterprising breeder who was faced with the ever-present problem of increasing the stock from the very small nucleus she had imported, decided to try using pale-colored Nubians. Two generations produced an astonishingly good animal with a passable fleece and no apparent kemp. I asked her why she chose that breed and she answered that the ears were more like the Angoras than any other available breed.

Colored Angora doe.

Texan Angora buck.

Australian Angoras have a grease-free fleece, unlike the Texans and some of the South Africans. This made shearing a problem until some way was discovered of keeping the shearing head cool. Wool from those with greasy fleeces is sold either clean or greasy at present, but as the crosses of the three types become widespread the differences will be minimized.

Cashmeres

Like the Angoras, Cashmeres were brought out to Australia in the 19th century and there still are large numbers of goats bearing those genes in the feral herd. Unlike the Angoras, where a small number of dedicated breeders kept a nucleus of purebred animals going, no one seems to have done the same for the Cashmeres. Our present herd has been built up from the feral source, and it is still worth considering ferals who show good Cashmere characteristics for upgrading. But as we go towards the turn of the century, the cashmere numbers are probably high enough for that course of action to be discontinued. The breed appears to be a sound diversification farm enterprise at present. However, in the future, many farmers will probably have changed to using Boers for this purpose as they are easier to manage.

Cashmere does.

Colored Cashmere buck.

Unlike Angoras, Cashmeres are closer to feral goats in their habits; this means that they bond with their kids almost automatically and consequently are much better mothers than the purer bred Angora, where bonding is a problem. Kidding losses in untended Angoras can be considerable (up to 50 percent are sometimes quoted).

White Cashmere is the most desirable commercially and originally the Chinese led the world in this fiber, but Australia is catching up. There is also a small and growing demand for colored fleece — mostly for hobby enterprises. If buying for commercial purposes, the micron count of the bucks is of crucial importance — provided, of course, the animals are of sound conformation.

Cashgora

This is the name given to the result of Angora/Cashmere crosses. In Australia it is not, at present, considered a marketable fiber, but in New Zealand there are quite a few breeders producing it.

In the United Kingdom during the latter part of the 19th century both Angora and Cashgora breeding was attempted, but neither proved a success possibly because the exacting nutritional requirements of large numbers of goats was not fully understood. Nowadays both Angora and Cashmere farming in the United Kingdom seems to range from the north of Scotland to the south of England apparently with success. It is perhaps too early to know whether the fleeces will develop a greater micron diameter than in the countries of origin which include the United States (Texas), Australia (via New Zealand) and South Africa. Some authorities consider the rich soils of the British Isles may not be conducive to the finer, more desirable micron diameters. This applies to Merinos and Alpacas as well.

Cashgora doe in a Cashgora and Cashmere herd.

Boer Goats

This breed has already been mentioned, it is now, eight years after the first edition of this book, becoming part of the show, commercial and breeder scene. It seems that they are, to a degree, prepotent; the first crosses are quite definitely Boers, even if slightly fluffier in the coat than they should be (when upgraded from fiber goats) and the color pattern, which is generally white bodies with black or brown around the head, appears to be very persistent. Of course the chestnut or red Boers are that color all over, but they have the same thick set and tough looking bodies. As mentioned, their docility is a particularly strong trait in the breed, even in the first generation, making them an ideal addition to the normal farming setup.

Boer goats.

Dairy Breeds

Nancy Lee Owen's very comprehensive *Illustrated Standard of the Dairy Goat (USA)* — which was, and may still be, a very useful publication for aspiring judges, is easy to understand and clear in its illustrations. The British Goat Society (BGS) also has a book on breed standards confined to Saanens, Toggenburgs, British Alpines and Nubians as do the Australians.

Saanen

Saanens are possibly the most numerous breed of milk goat in the world today, both for small holders and commercial milking enterprises. Saanens, as an unidentified white goat, have probably been in Australia since the importations of the first fleet. In the early part of the 20th century, the New South Wales Department of Agriculture ran a Saanen stud at the Nyngan Experimental Farm. Bucks and does were imported from France and Canada, the French ones mainly originating from Switzerland. In the early 1960s I had several Saanens heavily bred to the Nyngan lines, which were described as having long silky fringes. Few of these remain now and a strain of blue or wall-eyed Saanens also seems to have died out. The latter are reputed to be the descendants of goats that came from the German battleship *Emden* in the First World War.

Saanen buck.

Saanens that go back to the Nyngan lines are a smaller, more compact animal, often with long hair on the back legs and a fringe along the spine and around sides of the body. After World War II the new imports from the United Kingdom began arriving, which were all apparently British Saanens — a leggier, taller and more rangy looking goat. There were a number of very good goats brought to the Antipodes, quite enough to have materially improved the production and type in Australia. The herd recording figures for those goats, for both first and second years of lactation, are indeed impressive and we cannot come anywhere near them these days. There are now a number of studs breeding very good Saanens but type is not specified because on arrival here both types were designated as Saanens.

In the show ring there used to be a wide divergence to be seen in one class ranging from the stockier Nyngan (pure Saanen) types to the rather taller British Saanen, leaving the judge with the unenviable task of sorting them out. Both types had their very definite merits and there were obviously many mixtures between the two, but this explains why we often had such a lack of uniformity in this breed in Australia. The desirable height is set at 32 inches for does and 37 inches for bucks.

Saanens are a good breed for commercial farms because

Champion Saanen at show.

they are on the whole quieter and steadier than the other three breeds, with less tendency to jump. They do, however, burrow under fences, which need to be extremely well attached at the base. They do not seem to fret in feedlot or housed situations and are better on tethers. On the whole they are not as adventurous as their colored sisters, nor, some people say, as tough.

In Australia and anywhere with hot sun, it is most important that Saanens' skin, wherever visible, be a good tan color — otherwise they are liable to get badly sunburned or develop skin cancer or both. A tan skin is now mandatory in the breed standard in this country. In the United Kingdom many Saanens have pink or nearly white skins as there is not usually enough sun to burn them. As many of our earlier imports had pale skins, they still tend to crop up occasionally and careful breeding is necessary to maintain the tan skins.

Toggenburg

Toggenburgs originated in the Obertoggenburg district of Switzerland from where they were exported to the British Isles and the United States and finally they were exported to Australia.

They have tan to chocolate-colored coats (the darkness of the coat is often dictated by the copper intake in the feed)

Champion Toggenburg doe.

with the characteristic "Swiss markings," these they share with British Alpines. Neither of these breeds is supposed to have white marks on the flanks, and they are not registerable if these marks exceed the size of an inch in area. This is a rather uncertain dictum as the age is not specified, one inch on a kid will be double that or more on an adult. So it is better to make sure that your purchases are not carrying extra white marks if that rule is applicable. Height is 31 inches for does and 35 inches for bucks.

Toggenburgs tend to be more adventurous than Saanens and have a largely undeserved reputation for low butterfat — these are actually much the same as for Saanens. Like the Saanens, Australian Toggenburgs came from the United Kingdom, where there were, yet again, two breeds — pure Toggenburgs and British Toggenburgs. No one seems to have made it quite clear if they were all British Toggenburgs that came over, certainly they are the only ones that crop up in modern pedigrees — a long way back now. As with the Saanens, one can also be confronted with two distinct types in the ring, although in the main they appear to be like pure Toggenburgs.

Toggenburg doe.

Toggenburgs are a mentally and physically tough breed. Their jumping powers, if they have been allowed to develop them, are considerable. Fortunately, although not numerous, they have been in the hands of some very dedicated breeders and, of all the breeds presently in this country, they show the most uniformity. Like all colored goats their copper requirements are higher than white breeds and, if these are not met, bleached out coats are the consequence. However, these soon revert to the correct color when properly fed. I boarded a Toggenburg doe for six months who arrived looking very pale, when the owners came for her they did not recognize their beautiful chocolate-colored doe.

British Alpine and French Alpines

This is the third "Swiss" breed although, somewhat paradoxically, they are all French Alpine goats in origin with the markings being those of the Sundgau, which was evidently incorporated into the French Alpine breed in the past. They have been beset by inbreeding troubles for much of their existence. The first "British Alpine" was a doe called Sedgemere Faith, a Sundgau type, found in the Paris Zoo in 1906. Faith was a small goat with a grotesquely large udder and much more white on her body than we would be permitted on a British Alpine today. She was taken back to the United Kingdom to found a breed and even now many of the British Alpines in Australia go straight back to Faith in about 45 generations.

Over the next twenty years, stock descended from Faith, mostly through her very prepotent son, Prophet of Bashley, were initially crossed out to both Saanens and Toggenburgs in an attempt to increase the numbers and fix a type. Finally, in 1919, they were crossed with Nubians in an attempt to regain the real black color. By the 1930s the type was set enough to allow them to be admitted to the *British Herd Book*, and surprisingly — in view of the their diverse history — some very beautiful goats and consistent milkers were bred from then on. Much of the credit for this must go to Mrs. Abbey, who was an English livestock breeder supreme. Her Didgemere stud produced magnificent British Alpines for years and she did not hesitate to cross out to other breeds when she deemed it was necessary. This fact should be noted by breeders of British Alpines at all times because, with a breed whose origins are so limited, a reduction in the genetic pool can often manifest itself years later in an unfortunate manner. As I write this update we are in real danger of this happening. Most of our big studs are 80 percent inbred to the New Zealand imports of the 1950s (see below). As the herd book is closed as far as bucks are concerned, I feel out breeding may ultimately be our only answer. This unfortunate situation should not ever arise with the French Alpine breed. Milking animals are not productive if inbred, as has been proven many times in the cattle industry.

Champion British Alpine.

In 1959 the first registered British Alpines were imported to Australia. They were two half-sisters and two half-brothers who were related in the third generation. From this most inauspicious beginning we have managed to establish a breed at last. Eleven unrelated individuals two years running are supposed to be the absolute minimum number required to start a breed according to an excellent book, *Goat Keeping in the Tropics*, so it is indeed a miracle that the British Alpines have built up as much as the breed has. Probably the breed was saved by one fact alone, that the British Alpine type is carried as a recessive by many other goats, generally British Saanens and quite often a first cross will produce the desired type.

This breed seems to have a built-in long lactating factor, some individuals in the other breeds also have this ability but not all, as in the British Alpines. This factor makes them especially useful, often crossed with Saanens, as commercial milking goats because they do not have to be kidded fresh each year and each goat milks two years. The lactation factor appears to be dominant and in my experience crosses, whatever the color, carry it.

British Alpines in this country have evolved into a rangy, elegant type; the breed standard fixes their height at 32 inches for does and 36 inches for bucks which makes them the highest of the "Swiss" breeds. In the United Kingdom the British Saanens, Toggenburgs and Alpines are listed as being entirely similar, "the only difference being in the markings." In Australia, they are tall, independent and athletic and for this reason the breed is not suited to feedlot-type farming or to tethering. They are very intelligent and responsive to their handlers, often more so than other breeds and for this reason would probably herd well.

Due to their early adventures when the breed was being "fixed," we do still find a fairly wide divergence of types and we also find certain characteristics, particularly from the Anglo Nubian outcross, cropping up. One of these is their voices which are generally louder than Saanens and Toggenburgs and their often high hips (with a steep rump) and occasional Nubian-type straight back legs. Their ability to withstand high temperatures with little sign of stress is possibly another inheritance. On the

British Alpine kids.

hottest days they will be found in the sun grazing. When British Alpines were originally exported from the United Kingdom to Trinidad, these goats acclimatized the most successfully of the three breeds. Another possible legacy of their crossings with Anglo Nubians is a higher butterfat range than that usually associated with Saanens and Toggenburgs — in the region of four to six percent and above.

Nubians

The derivations of this breed are dealt with earlier in the chapter. They are a dual-purpose meat and milk goat in their countries of origin, but nowadays the milk is the quality most sought in the western world. This in-built meat-producing quality is useful when considering breeding meat goats because a cross on feral or any other breed usually produces a well fleshed, fast-maturing animal and, for this reason, it should be considered as a cross when upgrading Boers. A breeder in Western Australia used this factor with considerable success some years ago to breed meat goats. He removed the feral bucks and introduced grade Nubian bucks to a feral herd producing a very good product — due in part, of course, to the vigor of the hybrid offspring.

In Australia Nubians may be any color and marking except "Swiss" (as in Toggenburgs and British Alpines), and they produce some strikingly beautiful colored coats. This has been, in one way, unfortunate, making them into a "fanciers" breed, possibly to their detriment. Serious breeders have concentrated on their conformation and milking qualities, with the markings being considered an added bonus.

Nubian buck.

Like all goats from desert regions Nubians have long pendulous ears to keep the sand from their ears, eyes and faces. The nose has a high-arched bridge and their mouths were originally often undershot to enable them to browse high branches easily. This characteristic is not encouraged nowadays, as it can lead to an animal that cannot graze well on grass. It is best to breed to, and buy, an animal with a good mouth.

Their backs are not level as those of Swiss breeds are supposed to be, but rise gently towards the rump, often with a fairly pronounced dip in the middle. Their back legs should be straighter and sturdier looking than Swiss goats and the whole impression should be of a strong animal of good height.

Sell your books at World of Books!

Go to sell.worldofbooks.com and get an instant price quote. We even pay the shipping - see what your old books are worth today!

Inspected By: Nora_Cisneros

0009280650 3

As a breed they have rather loud voices, which can be a nuisance when they are in season or unhappy for any reason — especially in built-up areas. They have considerable jumping ability which should *never* be encouraged.

Notice the slight dip in the back of this Nubian doeling.

As mentioned in Chapter 1, in the 19th century Anglo Nubians were the best milking goat in the world, much prized for their large quantity of high quality milk and their extended lactations. Sadly these qualities, especially the latter, have been very largely lost. In the United Kingdom a *British Goat Society Journal* of 1990 blandly avers that they are incapable of 365-day lactations. Fortunately, in Australia there are a number of breeders with Nubians fully capable of this feat. However, they do have a partly deserved reputation for being poor milkers and having bad udders although dedicated breeders have worked hard to improve these points.

Unlike the British Alpines, there were a significant number — about 27 — of Nubians imported to Australia from the United Kingdom, which should have resulted in a good gene pool. Unfortunately there were a great many deaths very early on, and the breed was reduced to a number that dictated a fairly large degree of inbreeding.

Swiss breeds can be upgraded from each other and generally a fairly good sample of the desired breed can be obtained in one generation. But Nubians do not have this recessive factor, and three to four generations at least are needed to produce a good, true-to-type animal, which makes their upgrading from small numbers more of a problem than it was for, say, the British Alpines.

La Manchas

Their name suggests they are descendants of goats from Spain. There is no mention of anything like them in Africa from whence so many of our different non-Swiss types emanate.

They have what are termed "gopher" ears which are about one inch in length and possess little, if any, cartilage. They stand about 26 inches at the withers, weigh about 130 pounds and have straight, neat faces.

La Mancha breed.

Crossbreeds

Many commercial milking units have a fair selection of crossbred animals, tougher and more viable, and some do not conform to any of the above types. In Australia they can be included in the Identification Register. These goats may be tattooed and registered in the normal way and can obtain their milk qualifications. Sometimes they are bred this way on purpose, to make full use of the hybrid vigor factor, but more usually they are a first cross that has not thrown true to either parent.

Here is an example of a cross between Cashmere and Nubian.

The all blacks that crop up occasionally in the British Alpine and Saanen breeds, and are indeed an accepted type in the French Alpine breed, do occur — often when least expected. By 1997 there was a definite move to found these as a breed and a supplementary register has been opened. They are not really like any of the Swiss breeds and their high copper requirement was often a death warrant as few breeders realized how much they needed. I bred a few of them pure and in conformation they seemed more like black Saanens than anything else, despite the fact that mine were recessives from British Alpines (which were, and still are heavily crossed in the United Kingdom with Saanens when necessary).

Sometimes crossbreeding is undertaken with the purpose of "fixing" certain characteristics, and when the progeny does not conform to either parent it can be registered in the Identification Register, thus ensuring continuity in a bloodline. Crossbreeding should be undertaken seriously with breeders using only the best individuals of each breed. Too often people take two inferior goats of different breeds hoping to produce something rather good. Two inferior animals can only, in the normal order of things, reproduce their own inadequacies.

Chapter 6

Nutritional Requirements & Basic Feeding Practices

Availability of Minerals

A goat in its natural state is a browsing animal that finds its nutrients in leaves, twigs and bark with pasture as back-up. We often try to feed them from pasture as we do not have the woodlands they really like and these, even if completely balanced, would not begin to meet their specialized requirements. The chances of the paddock being really healthy are minimal. Not just in Australia, but in Europe now, too, paddocks are more likely to be completely out of balance and low on most essential minerals.

A browser, whatever the species, gets its extra nutrients from trees because their root systems go much deeper than pasture and their leaves contain many minerals that are no longer obtainable on the leached out ground surface.

A source of fodder much used in the United Kingdom (and possibly in the United States) in the past and hopefully still today is gathering the leaves of trees like oak, ash, apple, pear, plane, sycamore, etc., and storing them dry for winter fodder. It is not a commercial option as a rule, unless uncontaminated leaves in quantity can be collected, but it is certainly a good extra for the winter rations as well as being highly nutritious. Letting the goats have the leaves on a free-choice basis would be an excellent addition to the ration. If gathering them in country districts and roadsides make sure (ask the local officials) that sprays have not been used to control weeds on the roadsides; this happens in far too many places these days. If the leaves have been sprayed, they are unsuitable for fodder.

A study of the analyses in Chapter 2 will show the inadequacies of many soils. Few of us will be able to let our goats, fiber or otherwise, have unlimited access to trees and/or scrub except in situations where they are being used for scrub clearing. Even that is a finite resource and sooner or later the farmer has to resume feeding his animals from pasture.

Feed Requirements

Goats should be run on remineralized paddocks that are as healthy as possible, good mixed feeds with the correct calcium to magnesium ratios and a pH in the region of 5.5-6.0, which is somewhere near ideal. Clover pastures are *not* an option; in fact, clover will not take over in minerally balanced paddocks with the correct calcium to magnesium ratios. When these are right and the soil is healthy, there should be a mixture of grasses, plants like dandelion, plantain, etc., as well as some legumes — as many varieties as possible. Protected belts of browsing trees that the goats can reach, but not destroy, should be available as well. Care must be taken with fiber and meat goats that they do not become hooked up by their horns when trying to reach the leaves.

The lick on the next page is a good standard stock lick for fiber and meat goats. It is best offered as individual minerals free-choice and must be kept *dry*. A shelter of some kind is probably the best course because the type of feeder used for sheep would not be big enough for horned goats.

> Pat Coleby's Standard Stock Lick for Fiber and Meat Goats
> (Make Available Free-Choice)
>
> Dolomite
> Copper sulfate
> Fine yellow dusting sufur (99%)
> Seaweed meal

The only possible addition to this lick, where the cobalt levels are very low on the analysis, would be cobalt sulfate.

In districts where the quality of the pasture is such that bail feeding — even for milkers — is not necessary, the lick can be used on its own. These conditions do not arise in Australia; however, they do in United Kingdom and possibly in the United States.

Read the sections on the minerals to see why this mixture will cover all the basics. Only two extras will be needed; cider vinegar, which will have to be fed on land where the potassium is low, and cod liver oil, which must be fed in dry seasons or when paddock health is poor. The main reason for many of the deficiencies in the paddocks as shown on an analysis, and inevitably by the goats' ill health, is the extended use of artificial fertilizers or land which has been over-farmed and not looked after. Artificials tend to tie up or inhibit many major minerals, magnesium and sulfur in particular, and all trace minerals.

Copper

As will be seen from the section on that mineral, copper deficiency is one that *must* be met if the goats are to be kept in full health.

Feeding Prior to Kidding

All goats, even if not hand-fed, will need supplementary feeding prior to kidding. Weaned goatlings will also need a little extra, especially coming up to kidding; dairy goatlings will need about two thirds the milker's ration by the time they are due.

All does, whatever the breed, will need good feeding in the final two months — fiber and meat goats should have the lick mentioned previously available at all times and supplementary feed as well. For some strange reason many people do not feed pregnant animals — it happens right across the board. If this is not done the goats will "milk off their backs" as the saying goes, and their bones will become depleted of essential minerals in their attempts to provide enough milk for the kids.

For meat and fiber breeds, well-grown hay and some soaked barley (mixed 50/50 with chaff and bran) containing copper and cider vinegar to ensure easy kidding (see the section on potassium) should be provided in feeders spaced out enough to ensure all the goats receive a fair share. In Australia, sheep farmers used to spread the supplementary feed straight on the ground. This can lead to waste, worm egg contamination and, in very dry years, to the animals picking up too much soil and sand with their feed. These days there are many firms specializing in paddock feeding systems. Any feeder should be placed so that the goats have access from both sides, preferably under cover.

Prior to kidding, the supplementary feeding is best done in — or adjacent to — holding yards. This helps fiber and meat does become accustomed to being mustered and makes them easier to handle. It also makes them easier to segregate should it be necessary during or after kidding. This applies particularly to Angoras, who often have problems with bonding, and means they can be kept in and monitored until they are used to their kids.

Topping Up Meat Goats for Market

Most of the previous suggestions can be used to give meat goats supplementary feed. The amount of hand feeding, or lack of it, will depend on the quality of available pasture and the stage of growth at which the goats have to be topped up. If pastures have been analyzed and top dressed with the appropriate lime, sulfur, etc., little hand feeding will be necessary as the goats will have gotten what they need for growth in the pasture.

Dairy Goats

Chapter 2 suggests various systems of keeping dairy goats including grazing with supplementary feed, feedlot, yarding and so on. One cannot expect top production from a goat that is tethered unless the operator is very hard working and in constant attendance. The husbandry for dairy goats is quite unlike that for fiber or meat goats. A dairy goat's output is the greatest for its size of any animal — so to stay healthy and productive they must be supplied with all the minerals and food they need. Minerals can be added to the bail feeds on a twice daily basis — it is a practical and easy way to see that all nutritional needs are met.

Dairy goats will need very good pasture and, depending on the quality, one or two feeds of a concentrate mixture daily. Feeding in the bail is not recommended because some goats eat faster than others and it can also lead to pushing and shoving at milking times. It is better to have a system where the does can be fed either before or after milking and have at least 20 minutes to finish up.

I have used the ration below for 30 years and found it satisfactory. The goats have consistently produced milk of excellent quality and quantity.

Ration for Goats
1 part lucerne chaff
1 part oaten chaff
1 part bran or rolled wheat

half a part grain, preferably whole barley, soaked
seaweed meal should be provided ad lib at *all* times
a container of rock salt should be available, (but healthy, properly supplemented animals will rarely touch the latter)
dolomite, a dessertspoon (2.4 teaspoons) per head daily
sulfur, a teaspoon daily
(Both dolomite and sulfur can be mixed with the dry feed.)

This ration can be made up for two, 20 or 200 goats.

Trace minerals such as copper, cobalt, zinc, if very low on the analysis, can be added to the water that soaks the barley. This should be made up once every 24 hours and half added to the concentrates at each feed.

Unpasteurized cider vinegar at the rate of one cup to every twenty goats is also added to the water that soaks the barley. In 24 hours the grain will have absorbed all the water which should be exactly half in volume to the barley it has to cover.

When there are only one or two goats, a week's ration of barley can be soaked at a time; the copper ration must be worked out carefully to see that the goats get their full requirement; the barley will neither sprout nor go bad because the copper and cider vinegar prevent this.

Alternative extras, particularly in harsh conditions, not exceeding 10 percent of the feed, in order of preference are:

Extra barley — this is a top feed, it needs better land than oats, rye or triticale and also has the great advantage that it contains vitamin B5 naturally.
Sunflower seeds — no more than a tablespoon per goat. This is a very oily feed and can, when overfed, block the pores of the skin.
Flaked or kibbled maize — a tablespoon per goat. This grain contains only vitamin B6, and pigs in the

United Kingdom became very ill with necrotic enteritis when fed only maize on its own as there is no vitamin B5 in it at all.

Linseed meal — two tablespoons per head.

Any or all of these extras can be fed when conditions warrant it. Remember not to exceed 14 percent protein in any feed; high protein feeds have several disadvantages — ill goats for one thing. Also the goats will suffer mineral deprivation as the higher the protein, the higher the need for minerals, especially copper (see Robert (R.J.) Pickering's work on the subject).

Peanuts are not included in this list because they can be very unstable and produce deadly aflatoxins if not completely fresh.

Similarly cottonseed is not an option, it has to be heat treated to be safe and this is hardly ever done (Dr. Ray Biffin).

Goitrogenic Feeds

All legumes come under this category. They deplete iodine if they are eaten in excess — in extreme cases they cause the goiter to swell from lack of iodine, hence the term. More often the iodine shortfall is too low to notice, but a preponderance of buck kids will soon alert the goat farmer to the need for a more balanced diet. Doe kids need more iodine than bucks; in extreme cases the males are born alive and the females will be born dying and/or hairless.

The effects of goitrogenic feeds do not appear to be so serious in the United Kingdom as they are here, perhaps because the country is not iodine deficient — unlike Australia where the whole country is affected.

Irrigation alfalfa grown with artificial fertilizers is especially toxic. Lupines, beans (especially avoid soy), peas, of any kind and their hay, tagasaste trees and clovers. All of these fed in excess cause trouble.

Seaweed Meal

This is best supplied ad lib in the lounging shelter in very solid containers attached to the wall. These should be at a height where the goats cannot manure in them. Should the trace minerals be in reasonable supply in the pasture, this ad lib seaweed meal should supply all that is needed including iodine and selenium and only dolomite, copper and sulfur will be have to be added to the bail feed. If there is a disease outbreak, the mineral intake would have to be reviewed — healthy goats should not get diseases, (see Chapter 11).

Protein Requirements

As mentioned previously, one should aim for a feed with a protein content of about 12 percent and not above 14 percent. Well-grown barley can be in this range, depending on the degree of fertility of the paddock in which it was grown. Rolled wheat and bran, which is more easily obtained, should be about 12 percent but is often lower nowadays. Lupines are about 40 percent protein. Expensive and risky to feed, they should not exceed more than five percent of the feed, no more than a dessertspoon (soaked) per head daily; personally I have never fed them and am not convinced they are safe. Well-grown organic feed often reaches protein levels more like those that were found many years ago.

When we came to Australia in the late 1950s, proteins of up to 18 percent in grains were not considered abnormal. In 1996, I was told of a load of wheat which made six percent. As the paddock degeneration continues, this lower percentage seems to be the typical picture. Paddocks that have been analyzed, remineralized and properly farmed may yet again produce good value feedstuffs one day.

Milkers have a high requirement for carbohydrates and, no matter how much protein is supplied, if it is not backed up with a larger supply of carbohydrates the goats will become ill. Many people have disagreed with me on this and then, after diseases have plagued their milkers, have in desperation changed their feeding over to the mixture I suggest and found to their astonishment that the milk quality and quantity has improved. Only too often one reads accounts of some of the great milkers that used to be around, the catalogue of the ailments that afflicted them is horrendous. Mastitis, acetonemia, milk fever, even foot troubles, especially laminitis — all due to an unbalanced diet.

Bovine Spongiform Encephalitis

The "Mad Cow Syndrome" (BSE) in the United Kingdom, was caused by cows being fed reconstituted animal wastes or meal made from dead animals; they just got by when it was properly heat treated at high temperatures. Once the low heat method was started (around 1974 as far as I can ascertain) the trouble started. It is now illegal in Australia (as of May 1996) to feed any form of reconstituted animal waste to a herbivore. This pernicious practice only arose because the proteins as explained above were becoming lower and lower in feed. They could have used lupines, etc., and the animals then would at least have been safe, but the meat meal was considered the easier option — 270,000 farms in the United Kingdom were reported to have fed meat meal — this would have been unthinkable in the early days prior to World War II. We have seen the devastating and far-reaching results of this practice now.

Pasture

My milking herd was always run on the supplementary grazing system. It really does suit goats best, they have to exercise themselves when choosing where they wish to eat. If the pastures have been reclaimed and are minerally correct (fairly rare) they will obtain 70 to 80 percent of their feed requirements from that source. In Australia the protein content of pastures is generally the worst in the world;

the lack of calcium, magnesium and sulfur is the chief cause. In New Zealand, for example, according to information from Professor Max Merrall of Massey University, the normal protein content of the pastures is 25 percent at any time of year, rising to about 42 percent in the spring. Obviously, in these conditions little, if any, supplementary feeding is necessary, especially if good hay was fed ad lib. The picture on healthy farms in the United Kingdom is similar.

In Australia it is doubtful if pastures even reach 15 percent in the spring and in many cases they would not rise above nine to 10 percent at other times of the year. The shortfall *must* be made up by supplementary feed.

Weeds

So-called weeds, many of them classified as noxious — not because they are poisonous (some are), but because they grow rampant in ill-balanced pastures, are often much prized by goats and other stock. Blackberries, horehound, docks, hoary cress, Patterson's curse, St. John's wort, heliotrope and thistles to name a few — all safe if the goats receive their dolomite in some form. Fireweed (*Senecio spp.*) is *not* safe — ever, and only grows on very poor land indeed. A year after remineralization it is gone.

All the above will be rapidly cleaned out by goats on a new farm. Note: if the copper-bearing weeds, Patterson's curse, St. John's wort and heliotrope are very heavy, dolomite *must* be fed while the goats are on these pastures — it offsets the copper toxicity. While the "weeds" last, the goats will thrive on them. With fleece goats take care that blackberries, thistles, bathhurst bur, bindi eye and few others do not cause fleece damage or tangle them up — good fodder or not. Weeds are a sign of an unbalanced paddock when they grow in profusion. When a paddock is in good health 40 to 60 varieties of plants should be growing in the pasture, including a few so-called weeds and of course grasses and legumes.

Hay

Those 40 to 60 types of plants, if included in the paddock and hay, will mean very little ill health in any animal that grazes or eats from such a food source. Each plant has roots that go down to different depths and, in so doing, make available a broad spectrum of minerals.

Whole books have been written on making good quality hay, only too often well-grown hay is ruined at harvesting, often being left on the ground until all color is gone. This means all the vitamin A is lost. Hay that is allowed to get damp may mean mold at a later stage. Generally mold is due to sour, rotten growing conditions and will result in thiamin deficiency and may cause up to 250 different fungal diseases all of which can result in hormone damage and possible death.

Hay should come from a remineralized paddock, grown with no chemicals or sprays; mistrust hay that has only one variety of grass. Ad lib hay, as good a quality as possible, should always be available to milking goats, especially in damp areas. This can be supplied in racks made from weldmesh or chain-link wire, anything larger and they will waste and trample the hay. The rack should have a wire lid to prevent the goats pulling the hay out over the top and wasting it. Goats will very rarely eat anything once they have dropped it on the floor, however the resident horse or other animal usually does.

Hay for goats, like their pasture, should contain as many plants as possible; goats prefer a good mixture. Hay that is too rough for horses will often suit goats very well. It is wise if buying a new batch to try a bale on the goats first, it will save money if they do not like it.

A mixture of clover and grass, or first cut *dry-land* alfalfa with a lot of grass is good. Years ago I was offered some baled straw free for the goats' bedding as long as I collected it immediately. I took a load home thinking it would do for that purpose and to my amazement the goats ate it all. Closer examination showed that about half of it was a trailing weed that I (probably wrongly) called wire weed and they milked better on that "straw" than on "good" hay. Owen Dawson tells me that the weed is called wire, hog or knotweed — *Polygonum aviculare*, a native of Europe — which contains higher protein than alfalfa. It is not a native of Australia, but is mentioned in botanical books from the United Kingdom.

Goats are *very* fussy feeders, stale feed of any kind should never be left in troughs or racks, nor will they pick up feed they have dropped on the floor unless they are desperate.

Silage

The same rules apply to silage as to good hay — bad pasture could never make good silage. Nowadays this is being fed to goats, although my personal feeling (on reviewing results) is that it is not an ideal goat feed. Pit silage does not ever seem to have been successful for goats, occasionally one hears of farmers who have fed it without trouble, but there are far more accounts of disaster. I do know of several people who have fed the small round bail silage to goats successfully, provided it is correctly harvested — it should smell rather like dried figs when opened. This kind of silage is fed quite largely to commercial goats in the United Kingdom, and, provided there is some other feed available as well, they appear to do alright on it.

Pellets

After several attempts at feeding pellets given to me by feed companies to try, which the goats carefully sifted and spat out, and hearing of other people's disasters when forcing their goats to eat pellets, I've come to the conclusion that pellets are *not* good for goats. The reason seems to be that, having been hammer milled, they become mush in the gut — failing to provide enough fiber to trigger off the cud chewing response. This eventually ends up ruining their digestions. Anyone with doubts about the consistency of pellets should leave a handful of them in a bucket covered with water — the result looks like custard in most cases.

Another reason for not feeding pellets has surfaced in recent years. Sadly some of the firms supplying pellet makers with ingredients have not been too particular about the quality and there have been cases in the horse industry of animals dying from botulism and similar fungal diseases caused by these feeds.

Amounts of Feed Required

The ration fed to milkers and young stock will vary according to the time of year, this will be dictated by the quality (or lack) in the paddock feed. A quart measure of feed twice a day with all the minerals added as previously outlined is a reasonable quantity for ordinary milkers, provided really good hay and grazing is freely available. Milkers giving a gallon and above may well be given more, concentrates should be finished in about 20 minutes, if not, cut back the amount of feed, or find out why (anemia, worms, tooth trouble etc.).

The condition of a goat will tell you if it is being over or underfed. Feel the base of the goat's tail, if it is really fat, the animal is converting feed into fat not milk, so cut down the supply. Like all animals, some goats are better "doers" than others, which means that they make better use of their feed.

Top milkers cannot always eat enough in the first few months of the lactation to keep up their condition and produce milk so fall away as a consequence. As the lactation progresses, the milk will drop a little and they will be able to eat enough to keep themselves in good order. By the time they are running through they will look really well. Good milkers are never fat — think of good Jersey cows. Running through means that they are only kidded every two years and this seems to benefit both mothers and offspring.

Overfeeding

Many of the troubles in modern goat keeping seem to be due to overfeeding rather than underfeeding. Farmers feeding their goats to a budget have far less trouble than those able to spend what they like on feed. Unfortunately, many hobby goat owners are guilty of overfeeding their charges.

Feeding Large Numbers

In commercial setups where goats are fed in large numbers the manager must be very observant. The secret lies in sorting the goats into groups according to their feed requirements and habits. Weaker and/or younger does must be fed in one unit, away from the big, strong, older goats, who will quite often eat their own ration in double time and several of the weaker, younger goats' portions as well.

Both the big commercial farms I saw in the United Kingdom divided their goats up into units of 50; some were divided according to age, and others according to size. Each farmer will have to work out the best method to ensure that maximum production and optimum health are maintained.

Feeders and Hayracks

Goats are, without exception, the hardest animals on their equipment that I have ever worked with. They will either, quickly or slowly, depending on the initial strength, part welds (both in mesh and equipment), break troughs, pull hay racks down — all in equipment that one fondly imagined to have been erected for life. Therefore do not be tempted to put up anything temporary if it is for goats. All of their equipment needs to be as sturdy and long-lasting as possible. A design for a hayrack is shown on the next page.

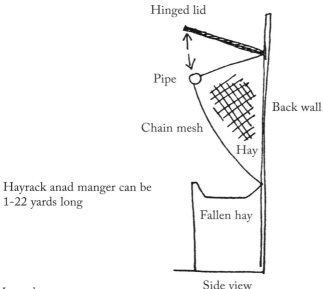

Hayrack.

Drought Feeding

Goats, like all stock, actually "do" better in dry weather provided one can meet their requirements in the way of fodder. During the really bad drought of 1983-1984, I had a farm with rocky outcrops covered in lichen, the goats ate it all, wearing their teeth down but keeping in excellent condition. In droughts the farmer will have to make do with feed that is available, particularly in commercial goat farming where cost and availability is the crucial factor. At these times imagination and a bit of extra work will often provide satisfactory answers. One sheep and goat farmer in New South Wales bought semitrailer loads of citrus waste from the processing factories (luckily within a reasonable distance) for about $11.00 a load. She had it spread across the bare paddocks and the sheep, angoras and dairy goats ate it all bit by bit, doing very well on it. Cannery wastes of other kinds can also be considered; in Italy they have been used as goat feed for many years.

At one stage of the drought the only hay I could buy was stripped phalaris from over 100 miles away. There was no way of seeing it in advance, but I had used phalaris hay (with the heads on) for goats several times quite successfully. The goats ate every stalk and the waste was minimal. Another successful standby was maize stalks after the cobs had been harvested. They arrived brown and rock hard, about eight feet long. At that stage I owned an old hand chaff cutter — very hard work — a mechanical one would do the job even better. I cut the corn stalks into one-inch lengths and fed it out with the bail feed. It was rather similar to sugar cane and the goats grew quite sleek on it, milking well. Sadly, when conditions improve most goat farmers go back to the easier option and forget about the feeds that helped bring them through droughts. These feeds often would be worth consideration in easier times. All those listed above were excellent forms of roughage, which is generally in short supply in drought. Nowadays when drought feeding I try to have a supply of very rough straw available as well as the good hay — it provides a balance and helps fill the gut.

After the Drought

There is a saying, which I earnestly hope now that we know the reason, will not be heard again. "Stock always die from unexplained causes for two years after a drought" — I had it said to me by farmers and vets when my goats (and many other farmers' sheep and cattle) were dying after the 1983-1984 drought.

Most of the areas in southern Australia that had been virtually bare of feed for 18 months grew only an African importation, capeweed (*Arctotheca calendula*) when the rains finally came; in other parts of the country, heliotrope was equally prevalent. The soil had been badly leached by the rains that preceded the drought — two months with eight inches of rain in each — which had helped lower the pH and, more importantly, leached the calcium and magnesium out of the soil. These plants like acid ground and no competition from grass (which does *not* like a lime mineral deficient soil and a low pH). The seed of capeweed in particular lives 14 or more years and just waits for the right conditions. It is a fairly high protein feed, and makes excellent hay — if it can be dried satisfactorily. The trouble starts for two reasons, capeweed causes a fatal magnesium deficiency very quickly and, in certain weather patterns such as cloud and drought, it stores lethal or sub-lethal amounts of nitrates. Some of the older farmers I spoke to later knew this, but a whole generation of farmers and vets did not.

Initially the goats, which were milking very well, started to scour. I found the only answer was to double up the dolomite in the feed and they settled down again. The same applied to sheep, as a couple of farmers who had lost large numbers found until I suggested the remedy. In due course the capeweed died back, a little grass started to reappear and it looked as though the worst was over. Three months later the goats, looking better than ever and milking well, started to die. The first one stricken looked much as though it were suffering from tetanus, the symptoms were similar, but the normal treatments had no effect at all. Finally the vet tried massive vitamin C therapy, but it was too late. In rapid succession my milking herd ceased to exist. Fifteen leased animals were hastily sent home when the second goat died, and 15 more of my own succumbed. The agricultural department vets were totally mystified; the suggestion that it was caused by not vaccinating did not work either, half had been vaccinated and they died even faster than the others. In some cases the cause of death was *Clostridium perfringens D* and in others, like the first death (only we did not know it at the time), classic nitrate poisoning.

In normal sunny weather the nitrate factor is not too serious, variegated thistle and other broad-leaved species as well as capeweed build up nitrate in their leaves. When the sun shines, an enzyme called nitrate reductase takes over and converts the nitrates into proteins and amino acids. However when the days are dull and cloudy (thunderstorms also accelerate the process) the nitrates turn into nitrites and the plant becomes very dangerous indeed. This process is triggered by excessively dry *or* wet conditions. The magnesium suppression came first as already explained. Next comes an enormous need for iodine, the nitrates (nitrites by that time) cause thyroid disfunction, this in turn causes the adrenal glands to cease working and totally suppresses the uptake and conversion of vitamins A, D, E and C.

We were all at our wits end when a friend (Michel Porcher of organic gardening fame) sent me three photocopies, two from a French magazine and one from Professor Selwyn Everist's book, *Poisonous Plants of Australia*. There it all was, as explained above. I started to feed iodine — in the form of two drops of Lugol's solution a day for one week, then on to ad lib seaweed meal and the deaths stopped as though I'd turned a switch. I had not been feeding seaweed meal to the adults because the drought had made me cut costs to a minimum, I could only spare enough for the youngsters — none of which died. Goats, unlike other stock engaged in primary production, were not eligible for drought subsidies.

Nitrate poisoning can normally be detected by an examination of the blood, which becomes almost black in appearance. All kinds of poisoning were suspected and the guts of the many dead goats were subjected to every sort of examination, but of course the blood held the key and it was not checked. Professor Everist makes the point that it is not good to spray capeweed with the usual recommended sprays, mentioning MCP and 2,4,5-T in particular, as they only exacerbate the fatal effects. He also stated that massive vitamin C administration was the only remedy — it did not work that time.

Capeweed can be controlled by analyzing the soil as mentioned in Chapter 2, and raising the lime minerals (calcium and/or magnesium and/or gypsum) to their correct levels and thus get the pH back in balance too. It also thrives on compacted soil which is why aerating is so important (the lower a soil is in calcium, the more compacted it becomes according to Neal Kinsey). This enables the grass and other plants to compete. Soil aeration also has this effect. When the capeweed is growing, slashing it several times to stop it coming into seed will also help because the goats would not find it so palatable. The year 1990 was also a bad year for capeweed. My last farm had been heavily dolomited (one ton to the acre), the goats had access to seaweed meal and were given Lugol's solution, a dessertspoon (2.4 teaspoons) between 25 goats once a week. It was enough to keep them out of trouble. Occasionally I raised the dolomite a little and gave a five-gram injection of vitamin C to anything scouring persistently, but that was all.

Heliotrope (*Heliotropium amplexicaule*) works differently, but the end effect is the same — death. This weed, which is a small plant with multiple pale mauve flowers, is very prevalent in some of the drier districts and is excessively high in copper. Goats and sheep can graze it extensively without ill results — until the rains come and the green feed comes away. Then the overload of copper in the liver causes a hepatic effect and the animals die of the "yellows" as it is known, due to the yellow appearance of the fat and organs at death. Having gleaned that information from the vet, I had to work out a way to keep my goats alive — they had no other grazing in early 1991. Dolomite is a natural antidote to excessive copper, so I made sure that my goats got their full ration. My neighbor who ran sheep put out dolomite licks for them with the same hope and none of our stock suffered any ill effects at all. The rest of the district did not fare so well.

Poisonous and Not So Poisonous Plants

This is a bewildering subject with many animals and particularly so with goats. Many so-called poisonous plants can be safe at certain times of year, and some apparently all the time.

Oleander is one which every authority agrees is lethal, there was even a case of people becoming ill at a barbecue because they impaled the meat on oleander twigs to eat it.

Yet I once saw a very healthy looking goat tethered on the median strip of a dual carriageway, happily eating an oleander. Quite obviously, judging by the state of the ones behind her, she was on her way down the row and had been for a while — the early ones looked decidedly goat-worn. I have seen a horse in the United Kingdom eating yew trees, which kill in seconds at certain times of the year, without any ill effects at all.

I will list poisonous plants alphabetically and mark them according to known toxicity or otherwise. Many of these are common to the whole world, a few are only known in Australia and others are peculiar to the United States.

Arum lilies — poison, goats avoid them.

Avocado foliage — palatable, but reduce the milk in lactating animals, causes mastitis if the eating continues.

Azalea — deadly.

Bitter almonds — contain prussic acid.

Black nightshade (*Solanum nigrum*) — see deadly nightshade.

Bracken — cumulative poison, but safe if there is plenty of other vegetation; only grows in humus-starved paddocks. Top dressing with manure (composted) from the goat house will totally discourage it.

Boobyalla — reputed poison, my goats ate it for years.

Capeweed — See this chapter and nitrate poisoning.

Cherry leaves — reputed poison.

Chokeberry (USA) — poison.

Cocklebur (USA) — poison.

Cottonseed — poison constituent gossypol, must be boiled to be safe.

Deadly nightshade — correctly black nightshade in Australia — goats avoid it. It is not the same, or as deadly, as the European variety.

Eucalyptus shoots — to be avoided — very high in prussic acid.

Fireweed — a poison that takes 18 months or so to act, see Chapter 11.

Green potatoes — cumulative poison, do not feed.

Heliotrope — high in copper, causes "yellows" in sheep, goats, see above in this chapter.
Laburnum — pods are deadly.
Larkspur — reputed poison.
Laurel — reputed poison.
Lilac — poisons the milk, but not the goat.
Linseed — can contain prussic acid, should be boiled for four hours to be safe.
Milkweed — poison.
Oleander — deadly in spite of the previous anecdote.
Patterson's curse — same as for heliotrope.
Peach leaves — poison when withered, best avoided.
Poke root (USA) — poison.
Potato haulm — cumulative poison, do not feed.
Privet — poisons the milk but not the adult goat, bad for young goats.
Ragwort — same as for heliotrope.
Red clover — reputed poison, possibly too high in copper for white goats.
Rhododendron — deadly.
Rhubarb — contains oxalic acid, leaves are poison, but I have seen goats eat them with no ill effects (they were on regular dolomite).
Soursob — same as for rhubarb.
Sorghum — not safe until over three-feet high, or in seed for preference.
St. John's wort — too high in copper for white goats.
Sudex grass — same as for sorghum.
Sugar gums — contain prussic acid, young shoots especially poisonous.
Variegated thistle — same as for capeweed.
Water hemlock (USA) — poison.
White snake root (USA) — poison.
Wisteria — doubtful, only feed very small amounts.
Yew trees — poison.

I would regard most garden ornamentals with suspicion unless proved otherwise.

Notes and Antidotes

St. John's wort, heliotrope, Patterson's curse and, to a lesser degree, red clover seem quite suitable for colored goats (if they'll eat them), but poisonous for white goats, who do not need so much copper. A teaspoon of dolomite powder tipped straight into the mouth — a film container is a good tool for tipping this into an upheld, open mouth — is an antidote to copper poisoning, so is vitamin B15 (obtainable from a vet). In conditions where animals may have to graze these plants, make sure that they receive supplementary dolomite.

Linseed is a great feed for goats, either raw, boiled or milled, but it must not be soaked. It can release prussic acid and cause poisoning similar to that of sugar gums, etc. Signs of this poisoning are foaming at the mouth and collapse — pharmaceutical chalk or dolomite given a tablespoon at a time in a drench brings about a speedy recovery.

Rhododendron, azalea and possibly oleander poisoning respond very quickly to vitamin C orally, about 10 grams for an adult.

Nitrate poisoning from plants has been covered earlier in this chapter.

Blood, Henderson, *et al*, have listed in *Veterinary Medicine* an extremely comprehensive list of poisons, some quite astonishing ones from the point of view of farmers, white clover is one. Perhaps it boils down to the fact that any feed in excess and on its own will cause trouble. Legumes, with their ability to deplete iodine, could certainly be regarded as poisonous but are in fact, when properly fed, very nourishing.

Everyone has a "poison" story, as mine previously with the oleander. Where I have put reputedly, it usually means I have seen those plants eaten by goats without trouble. Much depends on how and where goats eat them. If they have unlimited grazing and eat a little bracken or capeweed, they obviously know what they need. Goats in a confined situation will eat poisonous plants which they would probably avoid if on free range. Extreme care should be taken with yarded or tethered goats, they should only be offered safe food.

I have been astonished at the number of poisons vitamin C will detoxify. It was a blind try when it was first used on a friend's goats who were dying from rhododendron poisoning and had responded to none of the usual "cures." The owner rang me in desperation late at night, all the vitamin C she had in the house were some flavored chewable tablets, so we decided that she should grind up 20 for each goat. Next morning all were on their feet and grazing bar one, and she responded after a second dose.

Fodder Trees

This should really be part of the overall farm strategy: using fodder trees in shelter belts is an excellent standby. There are quite a few native Australian trees, some slow growing, others that grow as fast as the many non-indigenous species now available. Most European trees will grow in practically any part of Australia with a little care, water and attention; naturally they grow faster in areas of higher rainfall. It must be remembered that all trees like minerally balanced land and many of them like soils that are reasonably well drained as well.

A shelter belt should consist of slow-growing trees with faster growing ones on the outside. The latter can be available to the goats through a very well strained cattle boundary netting fence. The top strand will need to be reinforced as the goats will lean over it to get at the shoots they can reach. (A "hot" wire is an excellent deterrent). If the goats can browse from both sides of the fence, the shelter belt will need to be three trees wide. Unless a lot of electric fencing can be used, the trees should be planted a full yard from the border fence, anything less and the goats will reach their heads through and literally nip the shelter belt in the bud.

Slower-growing trees include:
European oak — all varieties.
European ash — all varieties.

Apple, pears and nuts — all edible for both goat and owner.

Kurrajong — a great drought standby.

Pine trees — all varieties, the needles dead or alive are very popular.

Macrocarpa cyprus trees — some people will disagree on this, but goats do very well on them.

Medium-growing trees include:

Casuarinas — all varieties, including she-oak, bull-oak; these trees stand trimming almost indefinitely.

Wattles (Mimosa) — all varieties, these have a limited life in dryer districts.

Elm trees — they sucker freely and are very nutritious.

Peppercorn trees — goats like this occasionally.

Poplar trees — all varieties, roots can run over forty yards underground.

Faster-growing trees are:

Tagasaste (Tree alfalfa) — *only* in well-drained, dry situations and totally allergic to glyphosate in *any* form.

Coprosma — also called mirror bush.

Willow — all varieites as long as the area is fairly wet, invasive roots (except for Chilean Willow), it grows fast from slips, and if layered will make a solid shelter wall, difficult to start in areas of high temperatures and hard frosts, but does well once established.

Paulownias — do not like to be too damp or too dry.

Mistletoe — grows abundantly in some districts and is another valuable goat feed; goats do urinate red water after eating it, but it contains a great many valuable minerals.

Trees like Tagasaste and some of the willows will benefit from fairly hard trimming, and if the goats cannot do it well enough through the fence the farmer will have to do some of the harvesting. Many of these trees will also produce a cash return if properly grown. Find a good book on agro-forestry.

Remember that Tagasastes are extremely sensitive to the glyphosate sprays, so much so that they can be used as an indicator of their presence. They will die very quickly indeed if glyphosate should be used near them.

Chapter 7
Psychological Needs of Goats

Many will doubt the need for a chapter on this subject, but 30 or more years in goat farming have convinced me that part of the success or failure in this branch of agriculture is acknowledging the fact that goats have psychological needs.

Leaders

Goats, like all animals, establish a pecking order which a wise goat keeper will recognize and use. The top doe should be first into the milking parlor; if not, she will make herself very unpleasant achieving that objective and may well hurt another doe in the process. Documented cases of leader does killing another one by crushing her against a wall or stanchion are not uncommon. It is better if the goat farmer takes the place of leader, in the shed at any rate, which should happen if the goats become accustomed to being led and handled. In the pasture there will always be a top doe, but in the shed it is easier if the farmer is in control.

All goats, particularly the older ones, should have names — ones that do not sound too similar. Goats soon learn to recognize their names when called or reprimanded. David Mackenzie, whose original book, *Goat Husbandry*, was full of excellent information, said if you get into a battle with a goat, make sure you win because otherwise you become a follower not a leader.

Bonding

Goats form very definite bonds, both with humans and their own kind. I know of two cases where goats, siblings reared together, who when separated pined and died in spite of all the best efforts of the vets and owners to save them. Veterinarians tell me they find colored goats are far tougher mentally and will fight more when sick than white ones. I would hesitate to make such a claim — having had all breeds at various times and noticed the same phenomenon — in case I was accused of bias.

Does bonding with their kids does *not* always come naturally, as breeders of Angoras have found to their cost. Highly bred milkers sometimes have the same trouble, but this may be due to the fact that commercial and show milkers are hardly ever allowed to suckle their kids. Others can be quite obsessive and will adopt any kid they can find in addition to their own. It is best *not* to allow this kind to bond with their kids at birth, or they will still be suckling them a year later. I once sold a doe three days after the birth of her kid, which I hand reared from the start so she would not see it and fret. Five years later I bought the doe back. On being released she walked right round the home paddock fence, then had a good look at the other goats. When I went out for evening feeding she was lying with her five-year-old daughter (whom she had never seen) and they became inseparable.

Bonding with Humans

Goats relate to people quite strongly and ignoring this fact is often the reason why commercial farms do not do well. The French goat farmers, who usually run about 200 animals, work it so that each batch of goats is mainly handled by one of the three or four staff (usually members of the family) and so learn to relate to one or, at the most, two people. In the long run I do not think large numbers of goats will do well unless some personal contact is maintained. (Cow people found this out years ago in the United Kingdom.)

Hand-Rearing Kids

For commercial farms it is better that all kids are hand reared, they will still receive their mothers' milk, but at the hands of the farmer. This makes the mothers settle into the milking routine without trouble. The kids then form a strong bond with the farmer, so they are more easily handled as adults, whatever the sex. If hand milking, this routine leads to much ear and face washing from the doe at milking time. David Mackenzie says that the doe thinks the farmer is her kid, and the kid thinks the farmer is its mother.

Goat's milk is best for kids if possible and it *must* come from CAE-free, *tested* does. If there is any doubt, the milk must be pasteurized first (see Chapter 13).

Replacements

If possible, replacements should be reared as a unit. They grow up together and will produce better throughout their milking life if they can be kept in the unit (a group of three to 10, or whatever). I used this practice for years with great success. The advent of CAE (see Chapter 11) upset the procedure, but now that we are coming out of that trauma the old routines are working again. Amalgamating different groups produced much unnecessary stress, with fights and bullying that often took too long to resolve. This undoubtedly upset the milk production in many cases.

With fleece or meat goats who do not hesitate to use their horns in these circumstances, the advantages of rearing all the weaners of each year together are obvious. Meat kids need not be removed from their dams, the object is to get them to a marketable size as early as possible and they definitely do better on their mothers.

It would, however, be wise to rear the meat does who will be providing the farm check with their young as well if possible. Many of them may be producing three kids in two years and feeding them for at least three months. The strain of kidding is far greater than the strain of milking — even for extended lactations. I learned this in an unfortunate manner; but even the worst experiences can teach us something. There had been an outbreak of mycoplasma pneumonia (also called pleuropneumonia) in the goat world during the sixties. In badly ventilated sheds it wiped out an astonishing number of animals. The veterinary profession were understandably getting jumpy about it and when they heard of my goats sniffling they arrived at the farm in short order and told me what to give the goats.

Tetracycline drugs were used (later on we found out that they were not effective against that disease, see Chapter 11). The goats were given the same amount of the drug that had habitually been given quite safely to sheep. Unfortunately no one knew that goats could only tolerate about a sixth of what a sheep could take. The result was mayhem, the first does to die were the pregnant ones who faded away three months into kidding, the milkers did not die for five or six months. They died from renal failure, acute anemia and bone marrow damage — all irreversible.

Free Range Goats

Goats, like sheep, always travel towards the prevailing wind; they seem to use this practice to maintain their sense of direction. Several times after sudden and complete wind changes, I had to ride out in the evening and lead the goats in. In each case they were standing on the highest ground they could find, circling, apparently in an effort to find a bearing. They were not using their visual powers at all. Once they heard me, they immediately followed me (horse and all) back to the main farm, almost with a look of relief.

Small and Single Goat Operations

For those who plan to have only one goat, be prepared to act as mother, sister, kid or whatever because, if you do not, the goat will probably call continuously for days. This is one reason, especially if living in a built-up area, why it is better to have two goats. If the two have not come from the same farm, keep them in adjacent yards for a few days until they are used to each other. Once they are accustomed to one another, they can be allowed together and the fighting will be minimal and mainly ritualistic. They will soon settle down.

Buck Psychology

A male goat in the wild establishes his supremacy, which he will hold until a younger, stronger male thrashes him in a fight and takes over. When this happens, the beaten buck either pines away and dies or goes off on his own hoping to attract the odd doe or two. In the farm situation this obviously must not arise, and stud bucks are far too valuable to be left in a situation where it could happen. I do know of one case where it did and a very valuable old buck pined away and died after being thrashed by the younger buck.

Out of the breeding season males will usually live quite happily together, but once the season starts, they must be separated if fighting starts. A very young and an old buck, provided they do not serve does in sight of one another, will often be quite contented together. As mentioned, buck runs should not be adjacent, but have an avenue between them. I ran seven stud bucks in this manner for years and it worked very well. But once when I had been out all day, I arrived home to find that one of the bucks had broken through the weld-mesh into the separating avenue. He had obviously been fighting for hours with the buck next door through the wire, both were covered in blood and totally exhausted. When I parted them, leading the offender back to his own repaired run, each flopped down to sleep for several hours — obviously much relieved to be able to rest at last.

Taking Does to Young Bucks

Does always have a marked preference for the large, mature bucks. If the mating is intended to be with one of the new juniors, on no account let the doe see the older (and more attractive) buck first because she may flatly refuse to allow the younger buck near her after this has happened.

Herd Bucks

In the herd situation a ratio of at least ten or twelve does for a buck should work well; actually bucks can manage to mate successfully with quite large numbers of does. I allowed my young milk bucks up to 18 does in their first season, but not more than one on the same day. It is probable that the quality of the semen is not as good on the second service.

In herds there should be enough does to stop bucks from fighting. Bucks will, of course, have to wear a raddle harness so that the farmer will know who has served what. They should not fight if the herd and paddock where they are kept is large enough.

Fright

When frightened, goats are totally unpredictable. They will jump enormous obstacles that they would not even consider negotiating usually and be totally unresponsive until the farmer can "get through" to them. This shows the necessity for the goats to be accustomed to the farmer. I was reminded of another factor when milking this morning, it was David Mackenzie's dictum that one should never sneeze in the shed. I was suddenly overtaken as I was milking and gave a loud sneeze — goats jumping in all directions was the result.

Chapter 8
Management

All Goats

Those who farm goats of all breeds will make their lives much easier if they establish routine management procedures. Goats, like most animals, are great creatures of habit and it usually takes from two days (for the bright ones) to a week to set a pattern.

Occasionally I have had groups of totally unhandled goats to board. They are put in a fair-sized yard with six-foot-high chain link or weld-mesh sides, where they are kept well fed and watered. Each time I take the food in, they usually hurl themselves against the farthest fence and stand in a quivering group while I fill the troughs. When this process eases off to just standing in a group — usually after three or four days, I let them out during the day. At night it will be quite easy to yard them without help provided I keep a "safe" distance from them — about four to five yards. They are carefully herded towards the open gate which, once reached, they will recognize and go through expecting to be fed. A couple of days later they will be waiting in the yard in the evenings.

Goats, like all stock, have a "safe" distance — this set by the amount of handling they have received. If you wish to catch one that has not had much handling, you must corner it and then get inside that invisible barrier. This is best done approaching very quietly with outstretched arms so they do not know which one to watch; if one is moving too fast the goat will take flight. It takes very good coordination to catch a collar, horn or leg in flight. If you have managed to continue the slow quiet approach (much to be preferred), stroke the animal gently all over as you catch it.

Restraints

A portable bail like the one pictured on the next page is an invaluable tool when handling hornless goats. It is made from welded pipe and is fully adjustable. It can be used for tattooing all ages, drenching, doing feet, washing or for any situation where restraint is necessary. It is easily made by anyone with an elementary knowledge of welding. The handling bail for horned goats is equally easy to make and needs a chain and clip to put across the top of the neck when in the bail. I first saw this on Frank and Liz Wroe's Angora property, and they find it invaluable.

There are also several systems for foot trimming and general handling of fiber goats and meat goats on the market. The farmer should go to an established farm and see what they use.

Rugging

All my milkers wear woven plastic sheep rugs in winter as it gets to 27 or 28 degrees Fahrenheit here at night. These rugs are lightweight and allow the skin to breathe while giving total protection from wind and rain. In very cold weather, they act rather like a wet suit by capturing a layer of air and keeping the animal warm. They are very reasonably priced and usually have a life of about two years. They can also be made from wool bale material.

My goats have worn woven plastic sheep rugs during the winter months for many years now. They take less notice of rain, frequently staying out to eat even when the

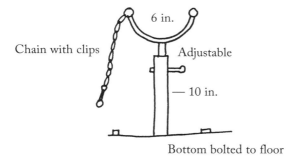

Utility handling bail for horned goats.

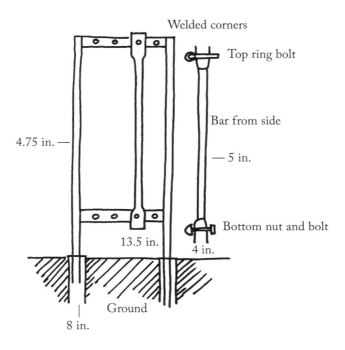

Utility handling bail for hornless or lightly horned dairy goats.

rain is quite heavy. Goats in heavy canvas horse-type rugs have been disadvantaged by them to the degree that their health has suffered. This type of rug becomes very stiff and heavy when wet. Rugging has been shown to improve annual butterfat levels by .07 percent in one herd where tests were carried out, they also save an appreciable amount of feed because extra feed is needed in the cold.

Back leg straps   Joined front at chest

Woven plastic sheep rug.

Clipping

If hand milking, clipping is necessary (with 30 or fewer goats, researchers have found that machines are hardly justified, either financially or in time used — one farmer in the United States milks 80 goats by hand!). Even with machines this would make for cleaner milking. The udder and back of the belly should be clipped at regular intervals — I usually clip the beards at the same time so when the beards grow I know I need to tidy up the udders. In the United States all goats seem to be clipped entirely in the summer months. I think there would be a lot of sunburned goats if that was tried here. From the aesthetic point of view, a clipped goat is not very beautiful so do not be tempted to try it for shows. I wrote that eight years ago and now I see far too many goats clipped for show; I can only assume they were not healthy enough for their coats to look good after a wash. In our variable climate I do not advocate it.

Feet

This is one of the most unpopular routines on a goat farm. Feet should be trimmed every six weeks, but often it is put off for much longer — until one realizes the goats are becoming uncomfortable.

If the farm has plenty of concrete and rocks, foot trimming can often by spun out to nine or 10 week intervals. In very dry weather the feet become extremely hard and it is better to take advantage of a spell of damp weather before getting the foot-rot shears out. There are also electric shears available now which would help fiber farmers with big numbers and the handling machinery mentioned above will take the back strain out as well.

Foot care also depends on nutrition; the correct minerals keep the feet growing evenly, and the right amount of copper in the ration precludes any risk of foot rot or foot scald (see Chapter 11). Chalky, fast-growing feet often denote too much protein in the diet and an attack of laminitis could be imminent.

The horn down the sides of the feet should be kept trimmed even with the sole although some goats grow too much sole as well, so this must also be cut back level. Occasionally, in very wet conditions, the outer horn separates from the foot and becomes packed with dirt. Cut an upward "V" shape into the affected part, and it will soon grow down. (See the drawing of feet in Chapter 4.)

Vaccinations

Farmers following a course of routine enterotoxemia (pulpy kidney) vaccinations should only use the two-in-one injection. This covers tetanus as well as "entero." There have been adverse reactions from the five-in-one which is not really necessary for goats; I've seen two cases of blackleg that were the indirect result of five-in-one vaccinations. The vaccinations should be done subcutaneously (under the skin) behind the elbow. This means that if the site "weeps" at all, it drains naturally although occasionally a lump is left at the site. Normally a doe is vaccinated four weeks before kidding which will protect the kids (in theory) until they are about four to five weeks old when they will have their first "shot." They are given a booster a little later and the vet will advise on further boosters.

I have never vaccinated my milking herd, except once under duress from vets, and several rather odd conditions resulted. This included an outbreak of CLA (Caseous Lymphadenitis) which started at the site of the injections. This was undoubtedly from a foreign organism gaining entry at the needle site. The vaccinations were done by vets. David Mackenzie, writer of *Goat Husbandry*, avers that properly fed goats do not contract entero and I have found him to be 100 percent right (see Chapter 6).

Older goats whose vaccination status is unknown should not be vaccinated. Once they reach the age of four or five years they will have built up a resistance to the clostridium concerned and may quite likely suffer (and die) from anaphylaxis if vaccinated. I have seen this happen, I warned the new owner not to do it, and I was unfortunately there when the goat died.

Whatever the cause, a sick goat nearly always dies from entero in the final analysis, but it is the result of the condition from which the goat is suffering, not the cause (see Chapter 11). Lately several vets have ventured the opinion that entero in goats is a much overrated disease and not nearly so widespread as is often claimed. Drs. Udzal and Kelly, speaking at a conference on goats in Queensland, stated that the pulpy kidney vaccination did not stop sudden attacks of entero, which I have known for years. They said that it could work with subclinical entero if a mob were in very poor condition.

Collars on Goats

As long as there are no horned goats in the herd, collars on all goats make handling much easier. It is not safe to use them if there are horned goats present because they can hook a collar with a horn, often done when rubbing against another goat, and a few twists later they are dragging a strangled goat around. If all goats wear collars, they very soon get used to being led or moved and are much more tractable as a consequence. Also, in very big herds, the collars can be tagged for instant identification and in areas where snakes are a problem, putting bells on some or all of the collars definitely cuts down the incidence of bites.

Ear Tags

I have found ear tags did not work very well on dairy goats, their ears appear to be too thick and those that I bought wearing them were well on the way to having septic ears. Fiber and meat goats, whose ears are thinner, do not seem to suffer.

Microchips

These could be used for animals that are to stay on the farm, obviously not for meat goats. The process is fairly expensive and means getting all the necessary microchip tagging equipment. Your vet will advise.

Buck Management

Bucks on commercial milking farms obviously must be segregated from the does. Animals that are "running through," that is, on their second year of lactation, cannot be mated indiscriminately. The bucks' runs are not only to keep the buck *in*, but also to keep enthusiastic does *out*; and the younger the doe, the more athletic and keen she is. Due to their very youth, they do not have a large paunch and udder to keep them on the ground and I have known them run up and over six-foot fences when in season. Fortunately when that happens they usually make no attempt to get out again, so at least the farmer knows that there will be a young and possibly immature goat to kid in due course. A five-month-old kid of mine did just that, and kidded at 10 months with two does and a big buck.

Her front legs started to bow with the weight and I put plaster of paris on to stop them from bending, removing it immediately when she was rid of the weight. She grew into a magnificent animal. Don't consider abortion — it leads to too many metabolic disturbances.

Buck Runs

These should be a minimum of 15 by 15 yards or larger. The runs should be rested at intervals and top dressed with dolomite or lime, whichever is needed, regularly. Two or three extra runs allow for resting — on this system bucks will be perfectly healthy indefinitely. It is preferable to be able to feed and water the bucks through the fence, especially in the breeding season, so if the farmer is absent they can be looked after by a relief worker. Bucks, like all entire males, can be peculiarly tempered, even with people they know, and must be handled quietly and firmly. This applies especially if there are does in season.

Like all goats, bucks thrive on human contact and in the breeding season a special garment should be kept for buck handling, so they can occasionally be given some affection. Unfortunately, their smell and general disreputable appearance at that time of year does not encourage sociability, but an effort should be made.

Houses

Each buck run should have a house with an all weather dry floor. This can be achieved by raising it well above the paddock with bricks or rocks or putting in a wooden deck. The house should be high enough to stand in, and the sketches in Chapter 2 show several cheap and practical shelters.

Gates

Gates on buck runs should always open inward and have a solid stop to prevent them being pushed outward. In a battle of pushing a 400 to 500-pound male will always win; they are very strong for their size.

Fences

These are probably best made from chain-link fencing. I used weld mesh for years, but finally changed to chain link when an old and very strong buck parted the welds by constant battering and walked through a new weld mesh fence in 24 hours. Chain link is also more resilient and does not assume the corrugated appearance of battered weld mesh so easily. If possible, a hot wire should be run around the inside at butting height of any buck run *before* the mayhem starts. I sold the buck who came through the weld-mesh later, and warned his new owners of his abilities — he was sold out of the breeding season, of course. They told me I was mad and that he was behaving impeccably behind a three-foot-high paling fence — they learned.

Summer Management

I try to have the bucks running together in the summer months. As long as there are no ovulating does around they usually will settle down to a quiet bachelor existence. The farmer will have to decide when to segregate them. Should a doe arrive out of the breeding season for mating, be sure to remove the required buck out of sight of the other goats. Let him serve the doe and then put him back at evening feed time — this usually stops any trouble.

Feed

Bucks, like the does, should have a hay ration every day. The run is mainly for exercise and they will not find very much grazing in it. Alfalfa hay is not desirable, it can lead to urinary and kidney calculi, so good grass hay is the best for them. Two feeds a day of concentrates should be given, which, in the breeding season, will be the same as the ration for the milking does. This amount should continue until the breeding season is well past and the bucks have come into good coat again and put on weight. Once this stage is reached the meals can be cut down, the object is to bring the buck into the next breeding season in top, but not obese, condition.

It is extremely important that each buck have a container for ad lib seaweed meal and that they receive cider vinegar in their ration at all times. Thus, even if they are on bore water, there will be no risk of calculi in the kidney, bladder or ureter. Of course, water should be available at all times. In summer months, the bucks should get a tablespoon of cod liver oil occasionally, and in drought this should be a regular feature of the diet. A real effort should be made to give bucks fresh green feed occasionally in the dry season as well. They go irreversibly infertile if they run out of vitamin A.

Leading

Bucks *must* be taught to lead when small and become accustomed to wearing a collar. When they are older it may be necessary to have them in leather head collars for extra control on public appearances. These can be made at home, or those sold for rams sometimes fit. Failing all else, a buck can always be led by his beard (if you can grab it), a process that

Leading a buck by the beard.

always leaves the leader scrubbing up slightly more frantically than usual, the smell always seems to be worse there.

De-scenting

Dehorning is explained in the kid section of this chapter. To de-scent the scent gland area, which is shown below, it must be cauterized at the same time as disbudding. The iron should only be held on long enough to burn the hair away. If the buck is polled, the scent glands will be just behind the poll "bumps." Personally I have never de-scented my own bucks, there have been cases of bucks treated thus being infertile, but it was never really established that the de-scenting was the cause.

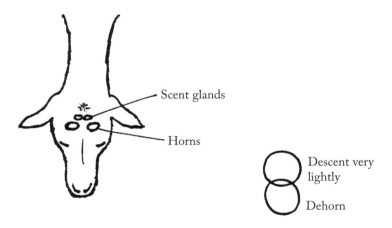

Descenting a buck.

Numbers of Matings

In theory this is limitless as bucks constantly spray themselves in the breeding season. However, I try to have a four-hour interval between matings if possible, or two does a day maximum, difficult sometimes when there is a rush of in-season does. Still, limiting the bucks to two does a day allows for more success in the long run.

Matings

A doe should be let into the buck run on a long rein (a lunging rein as used for horses is ideal) and pulled out once she is mated. I very rarely let a buck chase a doe around and never allow a young buck to do so, a doe could easily hurt him at that age.

With visiting does, if there is the slightest suspicion of a history of ill health or abortion, insist that the doe has a clear vaginal swab *first*, (this can only be done by a vet when the doe is in season). A doe can infect a buck when mating and render him sterile for life, or make him ill — it is not worth losing a buck for a service fee.

Kid Management

"Catching kids" will be covered in the first part of Chapter 11.

Birth

The goat farmer must decide *before* the kids are born which are to be kept, castrated, destroyed at birth, and so forth. These arrangements may well have to be reconsidered according to individual circumstances, but the general plan will be in place. This applies particularly to the family goat herd, where the children will have to realize that every kid cannot be kept.

Immediately after the kid is born the navel cord should disinfected with methylated spirit, iodine or alcohol. The kid, which should be strong and bright, must be checked for congenital defects. In particular check the vulva of the females (which should be like the sketch below) and scrota of the males — in which there should be two testicles. This is important whether the males are to be castrated or not. The absence of one testicle does not mean it does not exist, it may not have descended properly, in which case the "wether" will still be capable of serving does even if officially castrated.

Normal clitoris   Clitoris of intersex with bulbous development

Normal and abnormal clitoris (right) — the second type should not be kept.

Unthrifty and Underweight Kids

Dairy kids should weigh about five to six pounds at birth. Consistently lighter weights in the dairy breeds can mean that the viability of the herd is in doubt. Fiber kids are lighter — three to five pounds as a rule, and meat kids should weigh about four to 10 pounds. I have seen a dairy kid that weighed one pound and lived (it was a buck and castrated to become a pint-sized pet). Similarly I've seen kids that weighed 12 pounds and more, but the middle range is preferable.

Unthrifty kids should *not* be raised. Any kid that does not stand up unaided after about twenty minutes is suspect — unless there is a reason for its debility, like the mother being very ill. Kids that are born with the front legs bent over have contracted tendons, (see Chapter 11) which is a deficiency condition and will respond quickly to treatment. A kid that is weak for no very good reason should be allowed to die or be quietly dispatched, nature will have her reasons. Read the section on iodine in Chapter 9 in this context.

Like many others, I learned these things the hard way. Many hours were spent rearing kids on eye droppers, bolstered up by hot water bottles and so forth. It took some time to realize that the sickly kid grew up into the delicate adult which quite often died in its turn when trying to produce her own young. Sheep farmers have learned the same thing, abandoned lambs are very often left because they are unthrifty.

Early Kids

In my experience these kids are usually quite viable, but the health of the mother should come under scrutiny. The most probable cause will be lack of minerals or malnutrition in some other form. It can also be due to a doe being overly fat (another form of malnutrition). Kids that are born late do not denote any ill health in the dam like the early ones. This is more usual in a doe that is getting on in years, they will arrive in due course — and they are not always bucks.

Starting the Kid Off

If the mother is tested CAE-free, the kid may be left on her to suck and it should be up and looking for the teat without help very soon after it is born. Colostrum, which is the first milk, is high in antibodies and of a consistency that acts as a mild laxative. This ensures that the meconium (beastings was the old name), which is the first dark-colored manure, is passed straight away. The droppings then revert to a light fawn color as the milk is digested. The antibody action is only relevant to the doe's *own* kid, it cannot be transferred. In the section on CAE an alternative to the doe's own colostrum is described.

Whether the kid is to suckle its mother or not, it is a good idea to teach it to drink first. As each kid is born, take it from its mother and put it behind a wire door in the stall so the dam can see it. Milk out some of the colostrum, stand this in a bucket of hot water to warm up slightly and when the kid is about 15 to 20 minutes old, offer it a drink. Stand it up between one's calves then bring the bucket *up* to the kid's face, *do not* force the head down into it. If the kid has not suckled its mum it will drink readily. Give all the kids a drink like this to make sure each one receives its colostrum and learns to drink from a bucket. They can then be allowed back on mother if she is to suckle them.

This should be done in multiple births or the last kid may not get much colostrum.

If the kids are to be hand-fed — often the norm on commercial farms — go on from there. It is also useful for kids to learn to drink because if, for any reason, the mother cannot feed the kids, they will not forget what they have learned — this comes in handy sometimes.

Even when the does are feeding the kids, the udder *must* be checked morning and night and if there is any surplus she must be milked out. Often the kids suck unevenly in which case the full side must be milked out. It is also a good idea occasionally to take the kids away for a night. Then you will find out approximately how much milk they are getting — or maybe not getting, possibly due to old age or sickness in the doe.

Teats

Teats of both sexes should be two in number and have one hole in each. If there is any variation in this, the kid is best destroyed. Udder faults can be very serious in milkers and bucks. The latter can, and often do, pass on the fault. In registered goats these faults are a disqualification in either sex. This phenomenon is discussed in Chapter 4 and below.

Deformities in Teats

Two teats joined with a hole in each; sprig teats — with either one or more holes; teats with two holes at the bottom or at each side of one teat; blind teats with no holes at all — these are all deformities (see illustrations in Chapter 4). Kids with any of these must not be kept, they will in any case be virtually unmilkable. Tests have been done on milking does that have one or two double teats. Dye was put up one of the holes and has not shown up in the other hole on the same teat which means these animals have extra quarters. Unfortunately, when these does breed in their turn they can pass on a terrible deformity. A perfectly normal looking udder can turn out to have three quarters and only two teats, the blank quarter fills with milk as the other two do, but of course it cannot be extracted. Kids must *not* be kept with these deformities. Study the teat deformities chart in Chapter 4 for further information.

Supernumerary teats like those shown in the illustration in Chapter 4 are quite different. They *are* a breed disqualification, but if they are blind and a little distance from the main teat, they do not in any way interfere with normal milking. They should not be removed at birth because, if they have a canal, a weeping hole will remain and cause a multitude of problems. Once the doe is lactating and the teat is seen to be blind, they can be removed by tying a thread round the base of the extra teat and letting it drop off. Goats may not be registered with this deformity, but commercial outfits need not be concerned and it is not reason to destroy the animal.

Mouths

These must be checked to see that the "bite" is even. Goats do not have upper teeth in the front, and the bottom teeth should meet the pad squarely, as in the first illustration. Overshot or undershot jaws will mean that the kid will be unable to eat grass properly, so it should not be kept. All such deformities are breed disqualifications. Mouths are illustrated in Chapter 4.

Wry Face

See Chapter 4.

Mismarks

This applies to registered stock according to the rules of the breed. Saanens must be white all over, occasionally they may have a faint "biscuit" tinge. Toggenburgs and British Alpines must be brown and white and black and white respectively with "Swiss" markings. Very small white snips are allowed if they do not exceed the size of a dime in adults, but they are not desirable. Nubians may be any markings except Swiss.

Mismarked British Alpine kid born to two British Alpine types.

Horns

Kids must be checked for horns. This is done by placing two fingers above the eyes where the horns would grow and feeling if the skin moves freely or not. If the latter, horns are going to grow whether they can be felt or not, if the former the goat will be polled — no horns.

Horning is a matter of degree, some grow faster than others. Kids, especially bucks, may be born with horns that are quite protuberant, others will not develop them for two or three weeks — all should be disbudded by the time they are three-days old. In 35 years of dehorning a large number of goats, I have never known any ill effects from doing it at two to three days of age. Leaving them until they are older and stronger often makes the procedure quite traumatic for both operator and animal due to the effort of restraining them.

Polled goats often grow what look like light horns later in life, these are called scurs, which drop off periodically. This does mean that the goat has a very light degree of horning and can probably be mated with another polled goat safely. There is always an element of risk in mating two polled goats as they may produce an intersex kid.

Disbudding

The illustrations show two kinds of dehorning iron and one method of dehorning. These can be made from an electric soldering iron, or, as in the case of the bottom picture, from a 1.5 foot bolt with two nuts on the end. The handle needs to be long, otherwise when it is heated in the fire the operator will not be able to hold it.

I prefer to dehorn a kid standing up with as little restraint as possible because I find this method is less traumatic than holding them down on the ground. I stand with the kid's neck between my calves, holding it securely. I turn the head with the left hand (I am right handed) to one side, hold the ear out of the way, put my thumb on the horn area (beginners can clip the hair round the horn at first), and then with the red-hot iron held in the right hand, hold it firmly on the horn bud (having removed my thumb first). Once the hair is burned away, examine the area, and repeat the process holding the iron on for several seconds (some books say 10, but I find it is a matter of judgement) until there is a mahogany-colored circle over the horn bud. Turn the head the other way and repeat the process using a *fresh* red-hot iron for the second side.

The whole operation takes about 15 seconds for each horn and the kid will run away as though nothing has happened once the iron is withdrawn. In England it is now mandatory for kids to be dehorned with an anesthetic. The death rate from the latter is about 30 percent and up, and goat keepers who do not wish their kids to be injured (often irreparably) by an anesthetic at such an early age, have to break the law and dehorn their own kids.

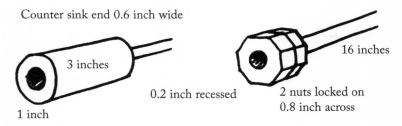

Dehorning irons.

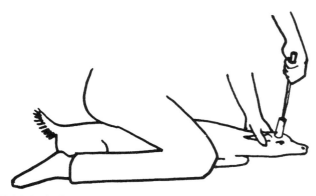

Fold the kid's legs and straddle it. With one hand, hold its ears back and press chin firmly on the floor. Apply hot iron.

Disbudding — alternative method (not preferred).

Please do not defer disbudding. In domestic and commercial milking herds, does with horns are dangerous both to the farmer and to the other does. They can cause very painful wounds, and a blow in the udder with a horn usually means the end of that doe's useful milking life. Dehorning adult goats is literally a bloody and painful process, extremely traumatic both for the animal and operator, even under anesthetic.

Occasionally I have left horns on kids which are to live in areas where there are marauding dogs, and which, when adult, must be able to defend themselves. In dairy situations — never do this, the damage that can be wrought by one horned goat is quite astonishing.

In days gone by caustic was used for burning out the horn buds — it is dangerous and cruel. I have seen several cases where the caustic paste became damp and ran into the eyes, causing pain and blindness. Nor should the horns be left to grow and an elastrator ring used, this can lead to horns that have dropped off and are only held by the nerves — this method is not humane either.

Castration

Male kids that are not needed for breeding must be castrated. The younger the animal is when it is done, the less pain and trauma. Even sheep farmers are realizing that the six-week marking is much harder on the lamb than doing it in the first week. I like to do it the first day if possible, all that is necessary is that you be able to hold two testicles.

There are three methods of doing this task. A good operator with a knife is possibly the most humane, the wound heals very quickly, although fly strike could be a problem. A bloodless castrator (illustrated) is my preferred method, but the male will be left with his libido in tact — useful if he is needed as a teaser, but not otherwise. The cam action of the implement cuts the cord of the testicles without breaking the skin.

The third, and most popular method (for the operator), is elastrator rings. The implement is shown below; the rings are put on *between* the body and the testicles, released and the testicles left to drop off. There is also an element of risk from fly strike in this method. Some years ago there was a survey reported in the farming paper, *The Weekly Times*, three groups of lambs were castrated, one group by each method. There were no adverse after-effects from the first two. When the third group was slaughtered, many of the lambs had bled internally — some had nearly two pints of blood in the abdomen. This is a possible explanation for the hunched-up appearance of newly marked animals. Again the younger the animal is when marked the less trauma.

Tattooing

At the time of writing this book microchip identification is being suggested, but at present the finer details have not been worked out and microchips will mean all show and goat clubs that run milk tests will need to carry the equipment to see the chip details. I feel that the present system is reasonably efficient and it does enable goats to be traced on a great many occasions.

This is an excellent lifetime identification if performed correctly. It is best done at six weeks of age, prior to that the

Elastrator shut

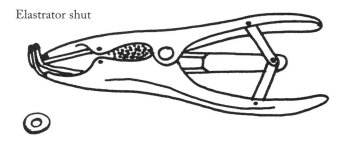

Elastrator open for application

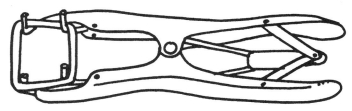

Bloodless castrator (sometimes called a Burdizzo)

Cam action

Castrating implements. Care must *be taken that the implement, whichever is used, is between the body and* two *testicles.*

tattoo tends to grow with the ear and becomes very difficult to read in an adult.

All registered goats must be tattooed — different countries have different codes and requirements. In Australia the stud letters are punched into the right ear and the number into the left by a special pair of pliers that holds the required symbols. The United Kingdom has a similar method with a modification that makes identification even easier. This is a different letter (announced by the British Goat Society) which is put in for each year, so the goat's age is obvious at a glance.

To tattoo — the utility bail shown earlier helps considerably — wipe the ears clean with methylated spirit or alcohol, otherwise the natural grease in the skin will stop the tattoo from "taking," then punch in the required symbols and immediately rub the tattoo ink into the marks. This order of operation is contrary to many recommendations, but means if a mistake has been made with the tattoo, the task can be redone a week or two later as a quick check is made before inking. At six weeks ears rarely bleed when tattooed, but for some unexplained reason when goats are older it often leads to mild bloodshed. If this happens, dry the blood up quickly as it tends to wash out the ink. *All* tattoo equipment should be kept spotlessly cleaned and disinfected between animals. I find that a small jar of water with dairy acid in it is excellent for this very necessary job — blood borne diseases like CAE can be spread this way.

Tattoo sites for dairy goats are found inside the ears.

Identification

Up to the age of six weeks, kids will have to be identified with an easy to see but difficult to remove (by the other kids) method. A Canadian goat farmer, whose kids ran into three figures annually, found cheap, light plastic chain and number tags fastened on with netting staples the most effective. For a small number of kids different colored collars can be used, but some sure method must be found.

Ear tags, as used in fiber goats, cannot be put on when very young, and they do not work well on dairy goats. Freeze branding could be used for colored ones, but tattooing as required by breed societies really seems to be the most effective, if not always the quickest to read. A flashlight behind the (cleaned) ear is often necessary to read a tattoo in older goats.

Leading

Time must be taken to teach kids to lead. This is even more important with kids left on their mothers, as they will not be naturally tractable like bucket-fed kids who will follow the person that feeds them. An adult goat that cannot be led and controlled makes the farmer's task extremely difficult. Usually two sessions, taking two or three kids at a time, leading them around is enough to give them the idea. Be careful to pull from the front so the pressure comes on the back of the neck, otherwise they collapse in a strangled heap on the ground, which is *not* the object of the exercise.

Of course any breeder who intends to show will make the leading training the first part of the routine, but it is just as important to be able to move the commercial milker around easily as her more exalted sisters.

Foxes and Other Predators

Meat and fiber kids in range conditions should be dabbed with a little Stockholm Tar on their rumps once the bonding is complete — *before* they and their dams are allowed out. This has the effect of masking their smell so the predators do not know they are around, yet it does not stop the mothers from recognizing their young.

Feeding Kids

In the *Australian Goat World* (the journal of the Dairy Goat Society of Australia) a very good article was reproduced from *Hoard's Dairyman* of October 2, 1986, first published in *United Caprine News* (USA) January 1986. It was on understanding kids' digestive systems. This article, which should be read by everyone who feeds baby goats, explains exactly why multiple feeds are not a good practice. Apparently the kid's digestive system needs at least five to six hours to digest a milk feed. If a kid does not drink its half-pint allowance in two minutes, remove the supply. Since reading the article, I have changed to two daily feeds after the first two or three days and found the kids grew as well if not better than before.

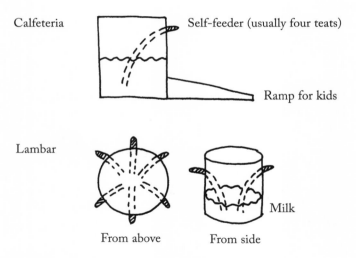

A Calfeteria and Lambar for feeding kids.

The article also explains that it is the slightly *under* fed kid that develops the best digestive system in later life. A doe whom I bought at six months was the daughter of a very big milker and was left on her dam. They were kept in rather confined conditions so that the doe was unable to escape from the kid as she would have done in nature, leaving it to play with the other youngsters for hours at a time.

Kid at Lambar.

This doe, like her mother, was an excellent milker, but her digestion was never good, due doubtless to the excessive amount of milk she obtained while growing. There is absolutely no doubt that too much milk when young is not good, the rumen fails to develop properly and the adult's digestion suffers as a consequence. Potential stud bucks who are going to be working in their first season may benefit from a quart a day for the first season. The section on bent leg in Chapter 11 explains how too much milk can totally alter the desirable calcium/magnesium to phosphorus ratio of the feed. This often results in deformed or soft bones.

All kids should have access to good grass hay and be offered a small concentrate ration with their minerals included daily. At first they jump in the feed and generally waste it, but very soon they will eat it with relish (if they are not being overfed on milk). Kids running with their mothers very soon learn to share her food and when they reach that stage, a separate bucket for them is a good idea. The ration should consist of chaffs and bran or rolled wheat, or the same concentrates that the milking does are receiving — just a smaller quantity.

Seaweed meal should be permanently available to kids and they should be allowed to help themselves. Kids are naturally curious and very soon learn to eat it. Seaweed provides natural iodine and trace minerals like selenium. A selenium deficiency is a potent killer of young stock and is becoming more widespread each year; the selenium in seaweed products is natural and safe. It should be remembered that selenium cannot be assimilated without sulfur (see section on that mineral in Chapter 9). Should extra selenium supplementation be required, consult the vet. Selenium quite rightly, due to its high toxicity, is not available to the general public in Australia.

It is important that dairy doe kids be taught to eat properly when young. "Picky" does never turn into good milkers. Does need to have a healthy appetite for their feed and eat up their concentrates well if they are to be useful milkers.

Milk Replacers

The ideal milk for kids is goat's milk. Replacers are probably poor economy in the long run. Naturally for those who do not have goat's milk available replacers will have to be used, but fresh cow's milk from a healthy cow is possibly the best alternative. Replacers, even when labelled for kids, should be treated with caution — make sure that those used are free of tallow, growth enhancers and antibiotics.

There is now a range of kid feeds that meet these criteria, which has only happened fairly recently. Tallow has an unfortunate habit of coating the kids interior so that it cannot absorb the necessary nutrients and the kid ultimately dies of bloat. A replacer could have cow or goat milk added for better results. If replacers are used with antibiotics or growth enhancers, the result may be satisfactory while the replacer is being fed, but disaster strikes when it is withdrawn. The immune system has not been allowed to develop normally and there have been many tales of illness and deaths following the use of such feeds.

Temporary Kidding Areas for Angoras

Angora farmers should set up temporary kidding pens similar to those used for sheep in the United Kingdom and for Angoras in South Africa. The animals close to parturition are brought into these pens at night and hand-fed in or near them until the young are born. When the lambs or kids are three to four days old and securely bonded with their dams, they are released into the herd, thus saving losses from predators and mislaid kids and lambs. The slight extra labor involved is considered to be more than offset by the kids that are saved.

Cashmere, Boer and feral goats do not have a bonding problem and should be able to kid safely in a paddock situation.

Temporary pens are found to be much more successful than special sheds, or the same area set aside each year. Diseases, particularly Caseous lymphadenitis (CLA), which is easily caught via the navel cord, are almost impossible to eradicate from soil and sheds. South African Angora farmers found that with the temporary yard system the incidence of CLA dropped almost to zero and was a far better control than vaccination (which has not been very successful). In South Africa the yards were made from thorn bushes (to discourage predators), in the United Kingdom they are made from bales, two on top of each other and the third laid crosswise on the top, making a slight shelter round the perimeter of the yard.

Goatlings

A goatling/doeling is a female goat that has not born a kid and is between the ages of 12 and 24 months. Because they are neither pregnant nor lactating (except in rare cases) and are virtually fully grown, the pressures on them are few, so they are the easiest animals to keep — they are also the easiest animal to get too fat. In commercial herds most will prefer to kid their does around 14 months and run them through. Only the kids born to the previous year's 14-month kidders will kid at 21 months — having been mated at 17 months.

Goatlings that are kept to kid at 24 months must be fed with care — otherwise they will become obese and learn to make fat not milk. In their natural state they would have kids by this time. Too much fat laid down at this point can cause metabolic problems with food conversion when they start lactating — and if fat is laid down in udder tissue, the damage may be lifelong. A fleshy udder generally cannot produce its full potential of milk. Often at shows goatlings with "bully" shoulders are seen, this is generally caused by the animal carrying too much fat between the shoulder blades and the body and is, of course, not desirable.

Goatling's diets must be carefully watched and if the paddock feed is too good, they will have to be moved to one which is less rich. They will still need a handful — literally — of concentrates night and morning. This is to ensure that they stay used to the routine of being handled twice a day and receive their necessary minerals. All are important, goatlings can contract mastitis, often unobserved, which can wreck their future milking careers. Seaweed meal should be fed on demand for *all* ages.

It is unwise to keep all goats to kid at 24 months as growth rates differ and very large, quick-growing animals should be kidded earlier. Frank Thebridge, one of the great goat keepers of the early days in Australia, ran the Rock Alpine stud which was justly famous for producing very high-class Saanen stock. One of his best known does was Rock Alpine Marietta (she carried a black gene and many British Alpines today have her blood). Mr. Thebridge told me that when Marietta was six months old he realized that if he did not kid her early she would become so enormous that he feared she might fail to breed later on. So he mated her to kid at twelve months old and she made record milk figures on her first lactation. She was, of course, not mated the following season, but run through for two years, kidding again when she was three years old. Occasionally, a single doe kid will be like Marietta and grow very fast — Frank Thebridge's experience and knowledge of breeding should be borne in mind.

Occasionally a goatling's udder will develop on one side only, this does *not* mean she will have a one-sided udder. Very rarely does the full side reach the stage where it will need milking, but if it does become tight, this will be necessary. Goatlings from high milking lines quite often come into milk at about a year old (possibly they should have been mated young). These young udders must be carefully watched and if they show the slightest sign becoming tight or distended, they must be milked out — once or twice daily if need be. The term for these early producers is *maiden milkers*. In the past many goat keepers seemed to be prejudiced against them, for reasons I never discovered, because they usually only come from high milking lines. If milking a goatling becomes a daily routine, it is often better to move her into the milking herd, let her settle down and be fed a milker's ration.

If the goatling is not milked when she becomes distended and uncomfortable, she may either take the law into her own hands and milk herself, or udder damage could ensue. Does that learn to suck themselves are a nuisance, especially as they rarely do it when people are watching and the farmer may not realize what is happening. The best way to stop self-sucking is to start at the head end, trying to render the udder out of bounds is difficult, if not impossible. Make a goat version of a horse "cradle." This is used to stop horses from tearing bandages off their legs. The animal can still eat, but not bend the neck. It used to be made from wooden one-inch dowels roped together, but with the goat it is easier made of leather as in the illustration below — an old leather army gaiter is ideal.

The same ration should be fed to all the goats, whatever sex or age, it is the amounts which vary. The minerals are constant; in the case of goatlings they will get them as

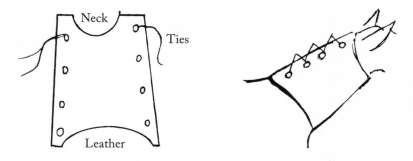

Restraint made from a piece of leather (an old army gaiter) to stop self-sucking.

described, but they *must not* receive the same amounts of concentrates as lactating does until the time mentioned above.

Does

A first lactation doe will have been fed as described above coming up to kidding. Sadly, there are still farmers who think it is not necessary to feed an animal until it is milking, in which case the doe will not have enough reserves to bring her through the heavy milking first part of her lactation. She will come into milk, perhaps quite well; but if she is a high milker, she will not be able to eat enough food to cover her production. This is where the reserves built-up before kidding take over. If she was not given correct rations, she will lose condition and milk and ultimately produce far less that she should have done (if she lives).

The self-sucking mentioned previously is not confined to goatlings. If a doe comes in to be milked with a sudden unexplained drop in her milk and is in good health, suspect this cause and cure it as suggested. Sometimes in first kidders the milk increases so fast that they feel uncomfortable. They turn around (when lying down) to see why, and, if they have been reared on their mothers, will suddenly remember what teats are for.

With very high milkers it is sometimes necessary to milk three times day at first — a doe with a badly distended udder and milk dripping out is very uncomfortable. When this occurs her production will usually settle down at about four months into the lactation. At that stage she will be quite comfortable on two milkings a day.

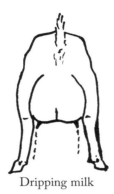

Dripping milk

Doe with over-distended udder.

First lactation milkers, especially if kidded at 14 months, *must* be run through; this means they will not be kidded the following breeding season. The production in the first year will not always be as high as it would have been had she been kidded at two years. Against that she will have been producing steadily for the intervening ten months instead of eating her head off doing nothing — definitely sounder commercially and mentally as well — scatty doelings can be a pain. By the second year, when the doe is running through, her milk will start to rise when the other does have their kids and often it will reach a higher level than when she came in originally. One of my best does, kidded at fourteen months, gave seven pints initially, slightly less as the summer warmed up, then when the autumn broke and green grass came she went up to the gallon, then dropped to about seven pints in the depth of winter. When shown at the Royal Show in the early spring (without kidding again), she gained her Q Star with 11 pints of milk and continued to produce well almost up to having her second kid the following year. This is the general pattern and, for both commercial and household milking, makes a lot of sense.

I feel this regime is the only answer, in this country at any rate, to commercial milking operations where it is not, I feel, feasible to have goats totally housed all year long as is done in Europe. There the ovulation period is controlled by lights as I saw done in the United Kingdom, but this can only be done when they are housed.

For years there was a prejudice against milking a mature doe up to kidding. Older does from good milking lines often do not dry up at all, or do so only for a week or two. Once the udder is full it *must* be milked normally, otherwise it becomes engorged (edema) with the sad result that the doe may never milk properly again. All too often one hears of does with huge udders which no one has milked coming up to kidding and having no milk when they do — lack of milking is the reason.

Saanen doe with kid.

Chapter 9

Minerals: Their Uses & Deficiency Signs

All necessary minerals should be present in the soil from which the goats obtain their fodder, whether hay or concentrates. Due to their browsing habits they will always need some extra, but the whole operation will work better if the paddocks are fully remineralized. Chapter 2 has shown that rarely, if ever, are the paddocks in full health, certainly not in Australia; the main minerals are either missing or they have been inhibited by too low or too high a pH or the ill-advised use of chemical fertilizers. No matter where you are goat farming, it is essential to know the full paddock analysis.

Without the correct minerals, vitamins cannot perform. The rough rule of thumb is that behind every vitamin deficiency there is a mineral one. Magnesium shortages lead to B vitamin deficiencies, except for B12 which is caused by a lack of cobalt, and so on. A magnesium shortage in the body can be induced by a fluoridated water supply, the use of too many artificials or land where it is inherently low.

Listed below are the main minerals. Lack, or very occasionally excesses, of these will make all the difference between success or failure in a goat enterprise.

Fiber and meat goats, whose requirements are not quite so high as those of milking goats, will be affected just as much by most shortfalls. A lack of sulfur and copper is critical in fleece quality and preventing internal and external parasites.

Four minerals and two vitamins are needed for fully healthy bones and teeth. Calcium and magnesium (as in dolomite); copper as in copper sulfate and copper-bearing plants; yellow dusting sulfur from supplementary feeding or in healthy remineralized soils; and boron, fed as borax (sodium borate) where necessary, but otherwise a constituent of seaweed meal; and vitamins A and D as in cod liver oil. Previous chapters have explained how to incorporate these materials in feeds.

Use as Basic a Form of Minerals as Possible

It appears that the simpler the form of minerals in the diet, the better they are assimilated. I saw a herd of cattle in the United Kingdom which were being fed a great deal of high-tech, mostly chelated, minerals of apparently synthetic origin. The cows were in a state of considerable ill health which was mainly caused by liver malfunction. Their bilirubin (liver function test) counts were below one and they should have been seven. I suggested that the farmer change their feed to a basic mixture we use here of dolomite, sulfur, copper and seaweed powder as outlined in this book and the improvement, which started in four days, has continued in all ages. A cow and goat dairyman in the north of the United Kingdom has had similar success and allows his milking goats the lick ingredients outlined in Chapter 6. The price comparison of the two regimes cheered both farmers up no end.

Proprietary Licks and Bullets — Effective or Not?

Various authorities have pointed out the uselessness of proprietary salt licks and blocks, as they never contain enough of the minerals that are really needed. On-farm supplementation, according to soil audits as suggested, is the most economical method for correcting problems.

A disturbing finding in several research papers recently is that bullets (boluses) for various mineral deficiencies do not always work, particularly in goats. Both bullet and scratcher (this is the piece of metal inserted at the same time as the bullet which periodically scratches it and releases some of its contents) have been found at post-mortem to be coated with calcium and therefore the contents have become unobtainable. Cobalt bullets have been mentioned especially. If you are using this method make certain, by taking blood counts if necessary, that they are working.

The Minerals

Boron (B) (Sodium borate)

Boron is extremely important in the cultivation of all legumes. Alfalfa will not grow or nodulate in a boron-deficient paddock (or calcium and magnesium-deficient paddocks). Roots will also fail if boron is too low; consult the soil analysis. In stock, this is an essential trace mineral needed in very small amounts. Calcium and magnesium will not be correctly utilized if it is missing, leading to problems associated with deficiencies of those minerals, like rickets and arthritis.

In areas that are mapped boron deficient, like the country around Bendigo in Victoria (Australia), arthritis in all stock is widespread. This is usually controlled by feeding either seaweed meal, which contains natural boron, or if that is not enough — as was the case in my milking herd — I had to add one gram per head a week of borax. Until this was done, the creaking joints were audible as the herd walked around the paddocks. Ten years later we find that the required dose is a teaspoon a week; sadly, soil deterioration seems to speed up as the years go by.

Calcium (Ca)

Calcium is required for the nervous and muscular systems, normal heart function and blood coagulation. It is also needed for bone growth.

However, calcium must *always* be considered in conjunction with magnesium — the two minerals interact and must be kept in balance at all times. An excess of calcium will cause magnesium to be depleted and vice versa. It is therefore unwise to feed calcium carbonate (ground limestone) or dicalcium phosphate (DCP) as these would cause a depletion of magnesium. Where magnesium is very high in the soil, the blood should be monitored to check if the levels are correct. If hand feeding, dolomite, which contains both minerals, should still be used because feed is frequently grown with "super" which ties up magnesium. On soils where the magnesium is high and the feed is well grown, fifty/fifty dolomite and ground limestone would be the best.

A headline in an English farming paper read: "Excess calcium gives cows mastitis." This statement, which could just as well have read "goats," is not strictly correct, but it made headlines in an English farming magazine some years ago. Calcium does not cause mastitis, but, by depleting magnesium which is needed to maintain udder health, an imbalance was created. Mastitis organisms were then able to gain entry and proliferate. Excess calcium in both the plant and animal world is linked with a weakening of the cell structure and lowering of immunity to disease, especially of viral origin. Dr. Neville Suttle, Moredun Research Institute, Edinburgh, also emphasizes "that feeding an excess of minerals like calcium can do more harm than good, predisposing to milk fever."

Calcium should be found in all feeds, alfalfa in particular if well grown (i.e., dry land alfalfa grown without artificials and with the correct minerals). However, its presence depends on two factors: 1) that the original soil where the feed was grown contained adequate levels of the mineral, and 2) whether or not artificial fertilizers were used — these reduce the levels of available minerals in the feed.

Conditions caused by lack of calcium are arthritis, uneven bone growth, knock knees, cow hocks, poor muscle tone (leading to prolapse), poor teeth, a general lack of well being and susceptibility to cold and, therefore, respiratory problems. Calcium and magnesium deficiencies will also cause lactation problems such as milk fever, mastitis and low milk production. Calcium in these situations is best supplied as dolomite, even in districts where the magnesium is reasonably high. For prolapse and associated disorders, the best type of calcium to use is calcium fluoride (nothing to do with sodium fluoride) which is obtainable as cell salts from most health shops. This disorder is fairly rare, so the remedy is quite economical to use. Goats who receive their dolomite regularly are unlikely to succumb to prolapse.

In Australia there are odd pockets where magnesium is higher than calcium, but they are rare. For permanent feeding, in the unlikely scenario that calcium on its own is needed, ground limestone (calcium carbonate) should be given. It is important to remember that calcium (and magnesium) assimilation depend on adequate boron, copper and vitamins A and D in the diet. See above regarding boron. Vitamins A and D (cod liver oil) should be available from sunlight and well-grown feed, otherwise they will have to be given as a drench or injection.

As has been noted in earlier chapters, calcium is essential in the ground in the correct ratios. Neal Kinsey notes that calcium helps the soil to flocculate freely and improve the texture. No crop or pasture will grow as it should if calcium (and magnesium) are too low.

Magnesium (Mg)

Magnesium deficiency appears to be rapidly becoming the biggest problem in modern conventional farming worldwide, possibly even in the United States where it is almost universally high (except for Florida and a few other mapped locations). Certainly in the United Kingdom magnesium deficiencies as we know them here did not occur before chemical fertilizers came into common use; now they appear to be rife in livestock of all kinds. Magnesium appears to be more readily inhibited than calcium by artificial fertilizers. Early experiments by Peter Bennet, an ecologist, indicated that one bag of superphosphate inhibited the uptake of five pounds of magnesium to the acre.

Considerable leaching of all minerals takes place in high rainfall areas. Both magnesium and calcium are rendered inert in the body by the sodium fluoride used in our water supplies. Goats should be given bore, dam or rain water and not water containing fluoride salts. Bores must, of course, be tested in case the nitrates are too high. Another little known cause of magnesium depletion is very high temperature. According to Dr. Harold Willis, hot weather (over 100 degrees F.) makes it unavailable in the soil and similar conditions certainly deplete it in the body. Magnesium is needed for all enzymes, both gut and muscle, to function correctly. I think this fact explains why so many farmers have told me that they felt dolomite had improved their feed conversion rate, as well as making mastitis and acetonemia in milking animals a thing of the past.

Seventy percent of ingested magnesium is needed for bone growth and the remaining 30 percent for neuromuscular transmission, muscular health and a healthy nervous system. The section on calcium shows how magnesium can be depressed by an excess of that mineral. It is also almost totally removed from the system by feeds high in nitrates, such as capeweed, variegated thistle and some broad-leaved plants such as clover. Animals grazing capeweed must have extra dolomite added to their ration while they are on it. Ruminants particularly, will succumb to scouring with general debility in the first instance and iodine deficiency in the second, followed by a fairly rapid death.

Conditions caused by a deficiency of magnesium (with calcium) include grass, lactation and travel tetanies, mastitis, acetonemia, arthritis, founder, warts (the virus that causes these prefers a magnesium-deficient host), uneven bone growth as well as most of the conditions related to calcium deficiency.

Animals whose bones have shown abnormal growth patterns or changes have been much improved (note anecdote in Chapter 4) in a few months by supplementation of dolomite. Deformed jaws, crooked legs, bone plates which have not knit as they should are generally due to a calcium/magnesium imbalance, very rarely are they congenital. From cases seen, it appears that it is seldom too late (or too early) to reverse it by properly supplemented feeding.

Excessively nervous behavior is also attributable to a lack of magnesium. Many animals have become much easier to manage when dolomite has been added to their rations, changing from excitable to quite calm individuals in a matter of hours or days. In my book, *Natural Cattle Care* there is a nasty account of a farm where magnesium had run out; fortunately it was easy to reverse, although expensive. Any condition that involves trembling, shaking, or excess excitability can nearly always be attributed to a lack of magnesium.

It should be remembered that overdosing with calcium and magnesium can lead to depletion of trace minerals and iodine. However, if oral copper poisoning is suspected, dolomite (with vitamin C) can be used quite successfully as an antidote.

Magnesium should normally be obtained from feed grown in soils containing adequate levels of the mineral. Unfortunately this is not always the case nowadays, as most feed is grown using artificial fertilizer and/or on depleted land.

In Australia, there is another source of calcium and magnesium for those lucky enough to live in the right areas — the artesian bore. Analyzing the water would be the best method of finding out what minerals a bore contains. They can vary immensely, even in the same district and can also these days, sadly, be contaminated with spray and nitrate residues.

Dolomitic lime, if decent quality, which means a ratio of about 45 ppm calcium to 20 ppm magnesium, is the safest way of providing these minerals, whether in the feed, licks or as a top dressing.

The following plants cause magnesium deficiencies in stock: Capeweed (*Arctotheca calendula*), and Smooth Cat's Ear (*Hypochaeris glabra*). They can be taken as an indication of a paddock's low status in magnesium and usually calcium as well.

Cobalt (Co)

Cobalt is needed for healthy bone development and for the health of red blood cells. Cobalt anemia causes persistent ill-thrift, depleted appetite, susceptibility to cold and death. In sheep, this condition is sometimes referred to as "sway back" or "coast disease," and goats are quite often affected as well.

The first sign of a shortfall is unhappiness, followed by lack of appetite,scouring, wasting and death in about 70 hours if nothing is done. The initial diagnosis, as soon as a goat is seen to be off color, is to take its temperature; feeling the ears and legs to see if they are cold will give you a pretty good idea that all is not well. If it is short of cobalt, temperature will be subnormal.

This highlights the importance of having the soil analyzed. Causes of cobalt deficiency are low pH and acidity, or occasionally, just to make things difficult, a high pH which inhibits the presence of the mineral and overuse of artificial fertilizers. I think it is uncommon to find an original deficiency. On a very poor farm where the low pH applied, after analyzing and top dressing with a ton of dolomite to the acre, the cobalt was back in the food chain in about 15 to 18 months.

Cobalt is synthesized into vitamin B12 in the gut and this synthesis can cease to work in cases of extra stress, illness or the administration of drugs. When this happens, the only way to reactivate the synthesis is to inject vitamin B12, orally will *not* work. This is a water-soluble injection which is completely safe, the body merely excretes any administered in excess of requirements.

Goats need two ml by intramuscular injection into the neck muscle, a very heavy buck of any breed could take three ml. Vitamin B12 is a great help in encouraging all ages to eat after illness or stress and should, in fact, be given as a routine measure on these occasions. It should also always be used for goats that have had multiple births to help the dam regain her full health.

Cobalt sulfate is extremely toxic (and expensive) and should only be given at a rate of 0.1 ounce (3 gr) between 40 goats per day; this is given by putting the sulfate into the water that soaks the barley in the ration. Extra amounts of cobalt sulfate should only be prescribed by a veterinarian.

Feeding seaweed meal ad lib or as a lick free-choice will help because seaweed contains all minerals in natural form. It is often enough to correct any problems. If the land is cobalt deficient, the best course is to have this amended by remineralizing (see this and previous chapters) with the required lime minerals. Consult Chapter 5 on soil deficiencies. Excess molybdenum will inhibit cobalt and copper. It is particularly important to make sure any analyst who tests the farm's soil includes cobalt. Copper deficiencies are serious, but far more so if cobalt is also missing. It is wise to use a firm that does not sell products as well as analyses.

Copper (Cu)

This is a mineral which can be top dressed in high pH areas where copper deficiencies are a big problem. An article in *Acres U.S.A.* some years ago told of farmers near Goyders Lagoon in northern South Australia in the 1940s. Their sheep had become virtually unproductive due to what we now know are deficiencies of copper affecting the wool and also causing lack of estrus. There was just enough copper in their systems to keep moderately healthy and that was all. The land (which was high pH) was top dressed with straight copper sulfate and the district never looked back.

I prefer to tell farmers to have the lick mentioned in Chapter 6 available at all times. Occasionally, in goats that are not used to hand feeding, some encouragement in the form of spreading a little grain on the top of the lick is necessary to get them started. They take what they need then, when they need it.

Neal Kinsey says that rust in crops and pastures occurs on copper-deficient soils, again a case of pulling the soil back into the correct balance so that the copper is available. Copper is inhibited when pH is either too low or too high; the latter effect is worse in droughts. André Voisin showed that nitrogenous fertilizers suppress copper. When I was investigating why copper shortfalls were so great, a researcher at Monash who was studying the mineral told me that copper is inhibited virtually 100 percent by superphosphate, so putting out super and copper is not an option — getting the soil into balance is.

I have covered the three so-called weeds that carry copper in Chapter 6. In the animal's body, copper is needed for optimum health, resistance to disease (especially of fungal origin), a healthy immune system and protection against internal parasites. Failure to come in season regularly is possibly the most uneconomic effect of low copper from a financial point of view. Goats whose copper levels are right cycle regularly at the correct time. In fact, when I first got that absolutely right, the bucks were run off their feet for a couple of days.

Worms, foot rot, cow pox, scabby mouth, orf, steely wool, ring worm, foot scald, proud flesh (which often occurs in bad cases of foot rot as well), Johne's disease, brucellosis (when deficiencies of iodine, manganese and cobalt are also present), poor fleece quality, animals that chew fences and bark, dark animals which are off-color, animals that are anemic or who are wormy are all suffering from lack of copper.

Anemia is a very serious effect of a copper shortfall, especially in Australia where most soils have adequate iron; without copper, iron cannot be assimilated. Given the fact that iron tonics are undesirable at the best of times due to iron depressing vitamin E, seeing that the goats get the correct amount of copper in their rations or licks is the obvious answer. Do not be led astray by the enormous amounts of iron tonics offered on sale in feed stores, they do not deal with the cause of the anemia, and as above, inhibit vitamin E.

The lick described in Chapter 6 is a list of essential minerals which has been tested and used for meat and fiber goats. I recommend that it be offered as individual ingredients free-choice and be kept dry. Goats are browsers by nature and therefore need more copper than other pasture eaters because they are used to the high levels of copper naturally occurring in foliage and branches. So they will take what they need.

Research carried out with humans in Japan in the 1960s established that black-haired people needed nearly six times more copper than fair-haired ones, and I have found those ratios to be right in dark and light goats. If there are any full black goats in the herd, it will be noticed that they often suffer from the above-mentioned conditions before their lighter colored counterparts. As long as dolomite is included in the ration, copper toxicity in goats does not seem to occur. I experimented feeding 0.1ounce (3 grams) per day to the entire milking herd (all colors) from February until November. I had the most trouble-free time ever. It was a really bad Gippsland winter and everyone was having terrible worm problems and we had nothing. In November I reduced the dose to a teaspoon per head per week and within six weeks the only all black goat showed very marked signs of worms. These disappeared after raising her copper ration again.

On the farm mentioned above in the cobalt section, there was 1,600 times too high iron levels in the soil. After two months the stock were trying to die of anemia and were riddled with liver fluke. Raising the copper in the ration cured the anemia and the fluke in ten days; in other words, they were then able to use as much of the iron as they needed. The drugs for fluke are very expensive and would, in any case, not have been practical as I was selling milk. Maintaining the copper at the correct level ensured there were no further infestations.

Goats whose coats are staring or look rough and fluffy in winter — or any other time — are always short on copper, and, if the hair is examined closely, a little curl will be seen at the end — another sure sign of copper deficiency. In real black animals, a red to rusty sheen on the coat is from the same cause. Of course many of these conditions can also be occasioned by bad worm infestations as well.

In *Hungerford's Diseases of Livestock*, he says that repeated and unexplained scouring is often caused by a lack of copper.

Occasionally goats will show the classic "spectacles" appearance when copper deficient; the skin around the eyes appear light colored and pulled away, making the animal look as though it is wearing spectacles.

Copper is highly toxic in excess and does not taste nice to humans at all, although I have spoken to number of people (of very fine physique) who regularly had it given to them as children (in cod liver oil). Yet, I have known of animals and birds who willingly ate it when they were badly deficient. The tolerance seems to be fairly high and cases of copper poisoning will not occur if supplementation is carefully monitored. In the section on CAE I give an example of goats who ate copper sulfate avidly when they needed it.

Zinc in excess suppresses copper. Too much copper kills, but too little does just the same. If copper is always fed in conjunction with dolomite it is usually safe. Copper injections should be avoided because an overdose cannot be treated. If too much copper is administered orally, dolomite, vitamin C powder orally and vitamin B15 injections together produce a very quick cure. Copper toxicity shows up as an acute liver attack. Copper sulfate *only* should be fed to stock, never copper carbonate. Copper carbonate is twice as strong as copper sulfate in the body and is not easily lost like a sulfate; therefore an overdose could easily be fatal. At the request of a copper researcher, I fed copper carbonate to my goats for four months, carefully halving the amounts. The exercise was not a success, the goats did much better on the sulfate variety of copper — and there is at least room for error if there was an overdose — there is none with the carbonate. Cancer and Johne's disease also denote a lack of copper.

Iodine (I)

The whole of Australia appears to be iodine deficient, surprisingly even in coastal areas. Iodine is not truly a mineral and thus cannot figure on an analysis. It is absolutely essential for the health of the thyroid gland, which controls the health of all the glands in the whole body — no thyroid — no life.

If a goat is iodine deficient, no matter what minerals or vitamins it is given, they will not be assimilated properly until the iodine requirements are met. In cold weather, the body's rate of iodine excretion is higher and it will be noticed that goats who have ad lib access to seaweed meal will take large amounts when there is a sudden burst of cold weather. Fortunately, the requirement is not very high and feeding seaweed on demand is usually enough to meet it. This is preferable to using inorganic iodine in the form of potassium iodine or Lugol's solution. Both are very toxic in excess as well as being expensive. A vet will advise on alternate forms of iodine supplementation if necessary.

If there is an iodine shortfall, obviously the animal's system cannot function properly and this should be considered as a possible basis of any problem. As seaweed supplementation has become the norm in all stock feeding worldwide nowadays, deficiencies should be unlikely; seaweed provides a wide spectrum of minerals in their most natural and assimilable form.

Signs of an iodine deficiency are, in severe cases, a swollen goiter and in mild form dandruff or scurf. It *must* be remembered that the signs of iodine excess and deficiency are identical, so, if a goat just brought in exhibits any of these signs make seaweed meal available on choice. Animals would not touch it if they have too much already in their systems — putting it in feed could lead to severe trouble.

A potent and common cause of iodine deficiency is overfeeding of legumes such as alfalfa, clovers, beans, peas, soy products, lucerne trees (Tagasaste), lupins, etc. These are termed goitrogenic feeds, because in extreme cases they cause the goiter to swell. Feeding too many legumes will cause a preponderance of buck kids. The doe fetus has the greatest need for iodine. If there is a deficiency, she does not develop and is probably reabsorbed. It is also possible in some cases that does are not even conceived. Occasionally, in iodine-deficient goats, strong males are born, but the females are born weak, dying and hairless.

Feeds high in nitrates like capeweed, when present in large amounts (such as after a drought), can also inhibit iodine and cause thyroid disfunction. If goats are on paddocks which are largely growing capeweed, iodine supplementation is necessary as well as extra dolomite (see magnesium section). Again seaweed meal would usually be enough. When I lost a whole milking herd due to capeweed, it was during the second year of drought and I believed I could not afford the seaweed meal except for the young goats. None of them died. It proved to be a very stupid economy.

Remember that an iodine deficiency should always be considered as the base cause of practically every problem. A blood test will show a large iodine deficiency, but may not always pick up a mild shortfall.

Iron (Fe)

In many soils in Australia iron is fairly plentiful to over supplied, due in part to the volcanic origin of much of the country. It is also often the only mineral left after extended use of artificials.

In the Gippsland farm, iron was 1,600 times too high and everything except the salt extremely low. This was partly due to the superphosphate and muriate of potash that had been used on it every year for a long time. Potassium and vitamin E are the major casualties as a rule and, in the body, iron destroys the latter which is why iron tonics must be avoided as far as possible. This situation can be amended quite easily by top dressing the area after a soil analysis. Bringing the calcium, magnesium, and sulfur up to the correct level is usually all that is needed.

The most important fact to remember about iron is that without copper it *cannot* be assimilated and many so-called iron deficiencies are merely due to a copper shortfall. Iron is necessary for the health of the red blood cells and, therefore, for the goats general well being. As mentioned in the section on copper, in spite of the prevalence of iron, anemia would be one of the biggest causes of sickness in stock.

Copper supplementation will raise iron levels very quickly and does not have the great disadvantage of totally suppressing vitamin E as supplementary iron does. Feeding fish meal is a good way to raise iron in the diet should there be problems; fish products are totally beneficial and, of course, safe (unlike meat derivatives).

Molybdenum (Mo)

This element is needed for maximum fertility. A deficiency can occur in land that is exhausted and low in organic matter, especially if the pH is below 5.0. However, high levels of the mineral are serious as it suppresses copper (and cobalt) utilization almost entirely. This is probably the reason why we were taught to adjust the levels of the other elements through top dressing and raising the lime minerals rather than altering the molybdenum.

Increasing the organic matter in the soil is the best remedy — in other words, get the farm onto a firm natural basis as soon as possible.

It should be noted that there are areas in Gippsland (Australia) close to heavy industry where the molybdenum levels are rising due to fallout from some of the industrial complexes. The vets who first drew my attention to this about 20 years ago were extremely worried about the situation.

Nitrogen (N)

Artificial nitrogen in its many forms is a substance that, as farming becomes more natural in character, will have to be phased out. Both Albrecht and Kinsey have discussed artificial nitrogens at length and say that when the system is fully organic, compost may be used successfully and artificial nitrogen should no longer be necessary.

Nitrogen is essential for plant health and it is found in every living cell; a plant is made up of two to eight percent nitrogen, and, in a good year on balanced soil, it obtains 70 percent of its requirements from the air. In a bad year, when conditions are less than ideal, a plant can only obtain 30 percent. (These are figures taken from Neal Kinsey's talk at Naracoorte, South Australia).

Rain and thunder storms (even dry ones) also bring extra nitrogen from the air and contribute quite materially, which is why people always remark that no matter how much they water, nothing is as effective as rain — it brings down nitrogen and an appreciable number of minerals as well. The amount of nitrogen delivered in thunderstorms can have a down side, especially on unbalanced soils. In the 1983-1984 drought, we had a long, dry winter with little if any sun and numbers of dry thunderstorms. Nitrate-bearing plants such as capeweed, variegated thistle, and any broad-leaved species like clover stored dangerous amounts of nitrates. An enzyme which is triggered by sunlight is needed to process this type of nitrogen into protein that stock can utilize. There was not enough sun to do it that winter and the subsequent spring. The nitrates turned to nitrites and hundreds of stock right across the drought belt died. The vets said that this always occurs after a drought, but I eventually found out (as related) that extra iodine would have averted the disaster. Years later, when talking to an earlier generation of farmers about this I was told: "Oh, we always knew you had to put out extra iodine when the droughts were bad." Information that never made the textbooks apparently, with unfortunate results.

In the United States the ground water, which is the chief source of drinking water, etc., is now no longer safe to drink due to the nitrates present. Adults can survive drinking it — just; apparently babies cannot. More seriously, the first reports of bores in Australia being contaminated are just starting to come in. The classic top dressing of super and ammonium nitrate or urea have been with us for most of the chemical era. Reports of urea poisoning have already been mentioned, animals can be saved if they are caught in time, but it is quite difficult to do. Artificial nitrogen has to be phased out.

Phosphorus (P)

This mineral is essential for healthy growth and life. In feed it should be kept in balance with calcium and magnesium, otherwise an excess of phosphorus will lead to bone fragility and many other problems of the kind associated with calcium and magnesium deficiencies.

Phosphorus will not be lacking in healthy, well-farmed soils where the minerals, organic matter and humus are in balance. Soils that have been heavily cropped are at most risk. But even they generally carry a "bank" of total unavailable phosphorus. This, as pointed out by William A. Albrecht and Neal Kinsey, can only be utilized when adequate calcium and magnesium are present.

Artificial phosphorus, as in superphosphate or any other kind of phosphorus, was originally used as a replacement for the animal manures that had been obtainable in the old world. There the stock were yarded over winter and the resulting manure was spread on the paddocks to grow the next crops *and* lime was spread in the autumn. This maintained the soil balance, because without the lime the phosphorus from the organic manure could not be fully used. Unfortunately that rule has been forgotten, until perhaps Albrecht, Neal Kinsey, *et al*, reminded us.

Now superphosphate has been used for years with no reference to the calcium and magnesium balance of the soils. This has lead to a situation where the excessive use of "super" has "locked up" a large range of minerals. These include magnesium, sulfur, copper, selenium, cobalt, boron, zinc and possibly manganese and molybdenum and quite definitely phosphorus itself.

The unavailability of these valuable minerals and trace minerals has meant a huge toll in animal health. Possibly the lack of magnesium and copper are the most damaging and reading the sections on these two minerals alone will make people realize the price we are paying for overuse of phosphatic fertilizers without referring to the important lime minerals at the same time.

Only since early 1996 have our soil analyses been showing total "locked up" phosphorus and the figures have been somewhat startling. Some dairy farms whose available phosphorus was about half to three quarters of what it should have been — 10 to 15 ppm instead of 20 ppm — have from 500 to 2,800 ppm "locked up." These "banks" will become available as the calcium/magnesium balance in the soils are corrected, so for the foreseeable future there is apparently plenty of phosphorus. I have not yet seen an analysis where the total unavailable phosphorus has been below 100 ppm.

This "bank" of phosphorus cannot be utilized until there is adequate calcium and magnesium in the soils — remineralization is the only answer according to Albrecht and Kinsey.

Phosphorus deficiencies in goats appear to be fairly rare; even when soil levels are low, the situation where the goats or any other stock are disadvantaged by the low levels does not seem to arise provided they all have access to the minerals as suggested.

Potassium (K, potash)

Potash is absolutely essential for all life and it is rapidly being lost worldwide due to chemical farming. In the Haughley Experiment it was proven that potash was the mineral most affected by artificials and, in spite of being replaced with muriate or sulfate (of potash), it was quite steadily lost. On land that was being farmed naturally, it was found to be self-renewing with no danger of running out. On poor, low-pH soils potassium is often inhibited by high iron as well as the acidity. According to Charlotte Auerbach's book, *The Science of Genetics*, a lack of potassium and vitamin C at conception can interfere with the true pattern of inheritance. So when breeding valuable goats, it is essential that the paddocks be as healthy as possible. The most serious result of a potassium deficiency is difficult births (dystokia) and having to pull every kid is *not* desirable.

In the short term, cider vinegar contains enough potassium to enable all goats to kid naturally, but raising the health of the paddocks is the best long-term solution. A vet from the Western District of Victoria told me that in the 1965 drought he had to pull 95 percent of all the calves in his area. There was no green feed and he considered a potassium deficiency was the cause. We know now that he was right.

A potassium deficiency causes constriction of small blood vessels in certain areas of the body. Dystokia seems to be caused by a lack of blood supply to the uterus and cervix in the final stages of pregnancy. In a normal healthy pregnancy the fetus moves constantly, until such time as it is presented in the birth canal in the appropriate position. The birth then takes place hassle free. When potassium is lacking, the fetus is locked in position and, speaking as one who has pulled an inordinate amount of other farmers calves and lambs, is virtually immovable. This is obviously damaging to both mother and offspring.

Selenium (Se)

There is absolutely no doubt that a shortfall of this mineral can affect goats very seriously indeed — to the point of death in a great many instances. When we first came to Australia in the late fifties, selenium deficiencies were only believed to occur in parts of the Western district of Victoria and around Canterbury in New Zealand. However, as time went on, rumors of selenium shortfalls in plenty of other places started to come in. In the middle 1980s a rather cross vet friend in Gippsland asked me to find out what I could about the mineral as he was sure many of the stock in his practice were suffering quite severely from a deficiency. He had been categorically told by the Department of Agriculture that a deficiency was impossible since there was no record of a lack of selenium in that part of Australia. Dr. Richard Passwater's book, *Selenium as Food and Medicine: What You Need to Know*, which I got from the United States, answered most of the questions and our growing knowledge of what low pH and artificial fertilizers were doing to minerals supplied the rest of the information.

Selenium, in particular, was at great risk because the sulfur levels in all soils were falling dramatically due to these same "artificials." Unfortunately, until the middle 1980s, the only firm that did a full soil analysis did not monitor sulfur, so we really did not know whether it had been there or not.

When I started keeping stock in 1960, I was told that all goats needed a teaspoon of yellow sulfur a week. This was to keep lice at bay — no one apparently knew that it was required for the amino acids. (Yet I have a booklet written by the CSIR in 1928 showing that this was so.) Within ten years that ration had become once a *day* for the same purpose. Warding off exterior parasites is only part of the function of sulfur, its amino acids are needed to assimilate selenium — which is more important. This was the information that the vets needed and it explained why selenium deficiencies had become so widespread. Not only was the selenium tied up, but so was the sulfur that was needed for its assimilation.

Like most trace minerals, selenium is equally dangerous in excess or deficiency. A small amount is needed for fertility, particularly of the male — without it the sperm tend to be weak and drop their tails. The mineral is also needed for healthy muscles. White muscle disease and muscular dystrophy are conditions exacerbated by a shortage of selenium. Complete and sudden cessation of growth, with or without muscle wasting signs, could be a sign of selenium deficiency in kids. I lost a few kids this way when I was first on the Gippsland farm previously mentioned.

Excess selenium can cause malformed fetus and/or poisoning. Sodium selenite (inorganic selenium) is only obtainable through a veterinary surgeon so accidental poisonings should not occur. However, Passwater's book indicated that one mg of organic selenium (as in seaweed meal) is equal to four mg of inorganic selenium — the seaweed is therefore the best method of supplementation.

Selenium is linked with vitamin E and often giving that vitamin alone will effect a cure in mild deficiency cases. In fact, this can be used as a test. If a sick goat responds to 2,000 units of vitamin E, you can be sure that it is suffering from a selenium deficiency. The farmer can then give seaweed ad lib as the lick suggested in Chapter 6; consult the vet for the best kind of supplementation if you need something beyond what is suggested here.

Sodium (Na)

Sodium must be in balance with potassium in the soil, but all too often it is in excess, especially if sodic fertilizers like artificial potash have been used. Real salt, that has not been purified, etc., should contain a fair spectrum of beneficial minerals and that, or rock salt, should always be available for goats at all times.

In excess, salt depresses potassium, causes edema (fluid retention), cancer and other degenerative conditions. Excess sodium also prevents the animal from using its fodder correctly. However, when goats have a craving for salt and eat large quantities of it, they are frequently looking for potassium — the two can interchange in the body (according to Louis Kervran). A farmer called me to complain that his sheep (on a very poor farm) were eating salt by the bag. I suggested that he obtain some cider vinegar and spray their hay with it once or twice a week. The salt consumption stopped within days.

There is generally enough salt for normal requirements in feeds. Goats needing extra are short of other minerals (see above). I found with goats that my animals who received their required minerals hardly, if ever, touched the salt on offer. However, when a new goat arrived it often spent the first few days punishing the salt blocks, but once the minerals from the feed came into play it, like mine, ignored the salt.

Sulfur (S)

Keratin, a very important constituent of wool, hair, hooves, horns, etc., depends on the cysteine in sulfur as well as adequate copper and it is only one of the many amino acids in the mineral. Protein levels depend on amino acids of sulfur, which may perhaps explain the falling proteins in crops as the sulfur in the soil runs out. Logically, it could be expected that the sulfuric acid used in the manufacture of superphosphate would mean a plentiful supply of sulfur — the reverse appears to be the case as mentioned in the section on selenium; sulfur levels are falling rapidly.

CSIRO did a short article on sulfur (*Rural Research Bulletin*, No. 22) and they found that without its amino acids the stock did not do well. They did not mention the onset of exterior parasites and may not have linked the two. They established that as far as oral administration was concerned, as long as the sulfur did not exceed two percent of all fodder taken in the day, it was quite safe. A very high margin.

Goats that are sulfur deficient may have lice or other exterior parasites. More importantly, they will not digest their feed properly or assimilate selenium because of the lack of amino acids, especially cysteine and methionine. Growing animals will not progress as well as they should if sulfur is missing. The mineral is often beneficial for skin ailments, both topically (applied to the skin) as well as added to the feed; therefore, skin troubles could be another sign of sulfur deficiency. Keratin is after all a component of the skin.

Sulfur can be added to fodder and fed to animals at any stage of development. For lice infestation, a heaped teaspoon every day for a goat until the lice go away will be necessary. A teaspoon per day is all that it needed for maintenance. Adequate sulfur will maintain animals in good health and free from "lodgers."

Zinc (Zn)

Zinc is usually fairly well supplied in soil; however, it is, as reported by *Acres U.S.A.*, another casualty of chemical farming so we may soon hear more about deficiencies than before.

Too high zinc can depress copper and this often occurs. The reverse, according to Robert Pickering, a researcher into copper of the James Lloyd Corporation in New Zealand, does not happen. Certainly the tests on my goats with very high levels of copper made no difference to their zinc requirements.

Zinc is necessary for a healthy reproductive system both in males and females, particularly the former as the prostate gland has a high zinc requirement. However, it is probable that there is not much difference between the requirements of the female reproductive system and the male, merely that the prostate problems are more obvious.

To date, a zinc requirement for goats beyond that supplied by seaweed meal has not arisen. Zinc sulfate can be used for supplementation, but the zinc in seaweed, like most minerals in their natural form, is more effective than using the sulfate. When sheep being shipped to the Middle East were dying soon after the boats left, it was discovered that it was a zinc/potassium shortfall. This was not remedied until seaweed was used as an additive, instead of the initially extra amounts of zinc and potassium added which had failed to remedy the situation.

Zinc is also indicated where recovery from sickness is not as fast as it should be, again seaweed products ad lib will probably be all that are needed.

Chapter 10

Vitamins & the Use of Herbal, Homeopathic & Natural Remedies

The Vitamins

Vitamins can be harmed by a number of factors. It is important to know that all vitamins are destroyed by light and by the use of mineral laxatives such as liquid paraffin; these should never be used under any circumstances. Olive oil or sunflower oil used for cooking are perfectly good laxatives which do not have this effect. Antibiotics can also destroy vitamins, especially vitamin K. Behind every vitamin deficiency there is usually a mineral shortfall and these will be discussed in the following section which examines the relationship between vitamins and minerals.

Vitamin A (Retinol)

Vitamin A is essential for the health of any animal. The health of the skin, eyes and reproductive tract — all depend on this vitamin. Deficiency signs can be a harsh staring coat, failure to conceive, reabsorption of the fetus, uterine ill health, death of the kids within nine days of birth, especially if deficiency is widespread, and ophthalmia (pink eye, sandy blight), genital and urinary tract infections, lack of resistance to interior parasites and, on occasion, cancer. Beta carotene is the precursor of vitamin A and can be used where small doses are needed. Like vitamin A, it is especially useful topically for eye problems (see Chapter 11).

Normally vitamin A is obtained from well-grown green stuff and is stored in the liver, thus ensuring a supply through the long, dry months of summer. But prolonged drought, interference due to hormone treatment, or electrical disturbances such as those mentioned below can all mean that the supply will not last the goat through the dry months. If the green feed is grown with chemicals, only 72 percent of the available minerals and vitamins will be present (research done in the United Kingdom). Well harvested, organically grown hay would be the ideal, but no matter how well grown, badly made hay with no green color is lacking in vitamin A.

All animals need a period of darkness every 24 hours, otherwise they cannot synthesize vitamin A properly. Occasionally this inability can be a inheritable factor and should a goat be consistently deficient in vitamin A, when others are not, it would be wise to cull her.

Other reasons for difficulty in synthesizing vitamin A are overhead power lines and hormone treatments. The latter have to be used occasionally and have their place, but farmers must take into account the fact that the goat will need extra vitamin A supplementation — regular dosing with cod liver oil or vitamin A and D emulsion would be the easiest way to give it to dairy stock. Intramuscular, high-potency injections can be used in herd situations (occasionally they set up sores at the injection site). Steroids have also been found to inhibit vitamin A, as well as assimilation of calcium and magnesium. However they should not figure in goat husbandry.

In *Goat Husbandry*, David Mackenzie's excellent work, he states that cod liver oil as such should not be given to goats as it can induce vitamin E deficiencies. But Mackenzie did not live in a country where these shortfalls are quite frequent; Australian goat keepers have often had to depend on cod liver oil as it was the only available supplement. It was certainly better than none at all and produced no ill effects, except possibly a slight lowering of butterfat levels just after administration — do not give it just before a milk test.

This vitamin is destroyed by light, it should always be marketed in opaque containers. Do not add it to drinking water where the light will destroy it. It is best put in feed that is going to be eaten immediately.

Vitamin B Complex

This is a range of B vitamins, all of which are needed for good health — lack of most B vitamins can be directly linked to low magnesium in the diet. Ideally, adequate amounts of all B vitamins should be obtained from well-grown feed, grains in particular (unmilled). In theory, parts of the B complex should only be given with a B complex supply, this applies to humans, but as animals normally obtain their B complex from their feed, this complex need not be given. Very few B vitamin deficiencies will arise if dolomite (for the magnesium) is fed regularly.

Vitamin B1, Thiamine

Vitamin B1 is a water-soluble injection, usually in 50 ml bottles. A deficiency of this vitamin can be very serious, signs are lethargy, staggering and lateral incoordination followed by blindness and death within 72 hours. B1 is destroyed by thiaminase which occurs in molds, either from moldy fodder (very dusty feed is often due to dried mold) or moldy paddock feed.

Vitamin B1 is obtainable as an intramuscular injection in 50 ml bottles from a chemist or feed store. Each ml of injectable B1 should contain 125 mg of thiamine. Assuming that this is so, doses of 6.6 mg to 11 mg per half pound of body weight should be given every six hours. Generally one dose brings relief but occasionally two or three are necessary.

Feeds very high in carbohydrates can increase the need for vitamin B1. In France where goats are heavily hand-fed, the addition of 60 mg of dietary thiamine daily is recommended. Occasionally an unexplained malaise that shows no clinical signs of B1 deficiency and does not respond to vitamin C or B12 will clear up quite fast when two to three cc of B1 is injected.

Vitamin B5, Calcium Pantothenate, Pantothenic Acid

Vitamin B5 is found in barley and if that grain is fed, whole and soaked as recommended, there should not be a shortfall. Vitamin B5 is needed for a healthy immune system and should be obtained from well-grown feed. Vitamin B5 and vitamin C together ensure that the adrenal glands function correctly — doing their task of maintaining the output of natural cortisone — which is very important.

Vitamin B5 is therefore a necessary part of the B complex. A deficiency can occur through poor nutrition. It is not obtainable as an injection at present. Should it be needed, 500 mg tablets (or lesser amounts) can be obtained from a pharmacy or health shop. This can be used as an added aid in combatting severe illness, 400 to 500 mg daily crushed up would be sufficient.

Vitamin B6, Pyridoxine

This part of the B complex is helpful against herpes infections. For those who show, vitamin B6 is a preventative against travel sickness (not necessarily tetany, that is a magnesium problem) and could be useful. About 250 mg of crushed up tablets in the feed the night before travelling is usually enough.

Vitamin B12, Cyanocobalamin

Both iron and cobalt are synthesized into vitamin B12 in the gut. In the case of a cobalt deficiency (check section on cobalt), it is necessary to restore the balance initially by intramuscular injection of B12, the oral route will *not* work until this has been done. This is because, in healthy animals, this synthesis is controlled by what has been called the intrinsic factor. In cases of deficiency, sickness or the administration of antibiotics, this factor ceases to work and must be reactivated by injected vitamin B12 before oral cobalt can be utilized. Vitamin B12 has a tonic effect on goats, from newborn kids that are off color to high-milking does that are a little low — all seem to respond to the vitamin.

David Mackenzie was a great advocate of this part of the vitamin B complex. He considered that its effects far outweighed the normal reasons for its administration and was hopeful that one day further research would be carried out on the vitamin (I hope so too).

Lack of copper (because it is needed to utilize iron) and cobalt can both be indirect causes of a vitamin B12 deficiency.

It should be a rule that if *any* drugs — especially antibiotics — have to administered, an intramuscular vitamin B12 injection is given at the same time. Many antibiotics and drugs seem to upset the gut flora and vitamin B12 helps to restore them. As this, like other injectable vitamin B and vitamin C, is water soluble, all three can, if necessary, be drawn up into the same syringe thus avoiding the necessity for three different injections. For this reason there is also no risk of an overdose, any excess is thrown off by the body.

Vitamin B15, Pangamic Acid

This is obtainable as an injection and occasionally in tablet form. It is a great help in restoring liver function after an illness or the administration of drugs. I would include one cc in the daily vitamin B injections of any sick animal for as long as necessary.

Vitamin C, Ascorbate, Ascorbic Acid, Sodium Ascorbate, Potassium Ascorbate, Calcium Ascorbate

Make sure when buying injectable vitamin C that it is 500 mg to a ml or two ml to a gram. It is important to make certain any supply is at this strength.

Any or all of the names above are used to denote various forms of vitamin C. This vitamin is synthesized in the liver of all animals (excepting cavies, some monkeys, passiform parrots and humans) from green feed especially, but all fodder (unless irradiated) should contribute to its production. Goats make approximately 10 to 15 grams a day.

Vitamin C is absolutely essential for life, the health of the collagen round the joints, spinal discs, the manufacture of cortisone (with vitamin B5) by the adrenal glands — all depend on this vitamin. Without vitamin C the animal could not maintain its health. Under stress of any kind — sickness, travelling, disease conditions and immunizations — vitamin C is depleted faster than it can be replaced (see Chapter 11 regarding immunizations). A heaped teaspoon a day is a good maintenance dose for a goat at risk.

At present, vitamin C is the only known substance that controls viral attacks. One well known dog vet (Wendell Bellfield in California) uses it in mega doses to cure viral diseases in dogs. He gives up to 200 grams on occasion. The vitamin is a great detoxifier; many poisons, especially those of animal origin like snake, spider and tick bites are all cured by vitamin C — the kind of snake is immaterial — a great advantage. Cancer is also controlled by large doses, either orally or by injection, but here one must bear in mind that because the animal makes its own supply as well, the cure is, so to speak, backed up and works surprisingly fast.

Ascorbate powder (or crushed up tablets in emergency) can be given by mouth; sodium ascorbate can be broken down with saline or distilled water for intravenous or intramuscular injections, or given by mouth. One heaped teaspoon of sodium ascorbate equals five grams. The only recorded effect of an overdose is diarrhea, and I have never seen it in a sick animal, in spite of some very high doses. For bad cases of scouring, possibly due to disease, oral vitamin C and dolomite mixed together often bring relief very quickly. Sodium ascorbate, which is virtually tasteless, can be mixed with milk when needed for kids.

Goats generally react quite badly for 30-40 seconds to intramuscular injections of vitamin C — if the goat shows no sign of pain at all, it is very sick. This is actually a good yardstick for recovery. When the animal starts to object, one can usually change to oral doses because it is starting to recover. Blackleg appears to be the only exception to this, the injection is very painful at any stage. In Chapter 11 amounts needed for various conditions will be given.

Vitamin D, Cholecalciferol

Vitamin D is synthesized on the skin from sunlight and is also found in well-grown feed. Light-skinned goats absorb it more readily than dark-skinned ones, which is why the latter do better in areas where the hours of strong sunlight are longest (they cannot get an overdose). Light-skinned goats can also be prone to cancer or heat exhaustion in tropical areas.

This vitamin is essential for bone growth and rickets is a deficiency disease due to lack of vitamin D which is also needed for the correct absorption of calcium and magnesium. In nature, vitamin D is always found bonded with vitamin A and on its own it can be highly toxic and is best avoided. Fish liver products, vitamin A, D and E emulsions or injections can all be used.

Feed that is too high in phosphates can depress vitamin D. Vitamin A and D preparations must be stored in light-proof containers.

Vitamin E, Tocopherol

This vitamin is essential because it plays a great role in healing, fertility and general good health. It is totally destroyed by excess iron, which is why iron tonics should only be used in an emergency and for a short term. Supplementary oral vitamin E is very expensive; horse suppliers usually handle the powder so it can only be used sparingly unless essential. Vitamin A, D and E as well as vitamin E on its own as injections are probably the cheapest way of administering it.

Vitamin E deficiencies should not really arise in goats, as it is found in all well-grown grains, especially wheat. But as it is destroyed by milling, old age and rancidity (which too often is not even apparent) the food source can be lacking. Raw, *fresh* wheat germ could be used in an emergency.

As mentioned in the section on selenium, the two are bonded in some way — selenium deficiencies and muscular dystrophy-type conditions all respond to vitamin E in the short term. The healing powers of vitamin E are particularly useful in cases of lung damage, where vitamin E can restore an accelerated breathing rate to almost near normal.

Vitamin H, Para-aminobenzoic Acid (PABA)

Until the middle 1980s this vitamin was referred to as a member of the B complex, but it has now been reclassified. PABA would not often be needed by the goat farmer unless an animal was suffering from, or likely to suffer from, sunburn — usually on the udder. It has great sunburn preventative properties and helps with the utilization of folic acid in the gut.

It is obtainable in tablet form, 200 to 500 mg could be crushed up and put in the feed — about 250 mg a day would be enough for a goat. PABA is also obtainable in sun-barrier creams which could be used on susceptible udders. This, of course, applies only to goats whose skin is not dark enough, the desirability for tan skins has already been stressed.

Vitamin K, Menadione

Vitamin K for coagulation of the blood is normally found in all green stuffs, particularly alfalfa. A deficiency should not arise in a normally fed goat. But, it is destroyed by irradiation and at the time of writing, irradiation of fodder is being suggested as a feasible preservative. Should this become the case, fodder thus treated should be avoided and the vet consulted if there is any likelihood of a vitamin K shortfall.

Using Herbal, Homeopathic & Natural Remedies

There are a number of excellent herbal books on the market — Juliette de Bairacli Levy's *Herbal Handbook for Farm and Stable* is one of the best. One must remember when reading her work that all of it was done in countries where herbs grew in the pasture and woods naturally and rainfall was on the whole regular and plentiful.

I turned to vitamins and minerals as they were always obtainable when animals were off color. Mrs. Grieves' *A Modern Herbal* is another book which I found useful because she often gives the mineral or other make-up of the herb and one can use this in place of the actual plant when the plant is unavailable. Homeopathy is now being used quite extensively and so are Bach Flower Remedies and acupuncture; there are good books on all three — study them. Look for a vet who uses *all* modalities including drugs occasionally when necessary, although this does not happen too often. Below are just a few of the items that I have used personally and successfully with animals.

Aloe Vera

Aloe is a plant of the succulent group that grows naturally in parts of Australia and the United States. Those lucky enough to have the actual plant often use the leaves directly, otherwise it is obtainable in liquid, ointment or gel form. Care should always be taken with any creams and mixtures from plants or other sources, always checking to see that they have not been scented or added to — the original is always best. Aloe vera can be fed or used externally, I tried the latter on a badly ulcerated wound I had in a buck goat that I bought with severe foot rot. The feet were easy enough to deal with, but the ulcer which was near the hock was of a long-standing and very obstinate nature. After trying anything and everything without success, the aloe vera effected healing in three days.

Aloe vera plant.

Apple Cider Vinegar

This simple and easily obtainable liquid is invaluable anywhere potassium is often found in short supply. It contains natural potassium in a safe form. It should always be bought in bulk and unpasteurized. Nowadays the demand is such that nearly all fodder stores sell unpasteurized cider vinegar in bulk and it is often grown without chemicals as well. The pasteurized variety is not popular with animals, nor is it so effective.

Feeding quantities of apples as such can lead to digestive problems in any stock, but they will tolerate cider vinegar in large amounts and it is wholly beneficial — a quickly assimilated source of potassium as well as other trace minerals. When I first read one of Dr. Jarvis' many books on cider vinegar, I did as he instructed and left a container for the animals to help themselves. It may have worked in Maine where the deficiencies were not so great, but I could not afford to continue it ad lib in Australia.

Cider vinegar maintains the correct pH in the body, which is probably one of the reasons it is so useful. Because of its potassium content, it is invaluable for all animals coming up to breeding. Potassium deficiencies cause blood vessel constriction, affecting the extremities and it seems, the cervix and uterus, in the final stages of pregnancy; dystokia is the result. I first used cider vinegar on my milking goat herd after a season of very difficult births. The next year I was amazed at the difference, even the largest kids from maiden does arrived relatively easily and in very good health. Many stock owners and human mothers have observed similar effects.

Cider vinegar helps prevent bruising and assists the tissues to recover from exertion. Given regularly to stud males, it will help prevent urinary calculi and this is especially useful if your male stock is limited to hard water — as is the case on many properties in Australia. Cider vinegar added to feed twice a week would be sufficient to stop stones in the urethra or kidneys and prevention is certainly better than cure for this dangerous condition. A dessertspoon (2.4 teaspoons) twice a week would be enough for most animals.

It can also be used as a mild cure for skin conditions like ringworm when it is too close to the eyes to use a copper wash; rubbing it in well two or three times a day for a couple of days is usually enough. Those wishing to learn more about cider vinegar should read any of Dr. Jarvis' very interesting little books on cider vinegar. There are various editions available.

Arnica Montana

This is a perennial herb that grows in the mountains of Europe and it is now being cultivated successfully in Australia and other countries. It is best used in homeopathic tinctures, pillules and ointments which are generally available in health shops — as always, make certain the product does not contain additives. In homeopathic form it is an excellent painkiller. I have used it postoperatively with astonishing results by normal rules. The dog concerned had no idea she'd had an operation and did not try to scratch or lick the site at all. It seems to have a healing effect as well as the dog in question has no scar from an operation to remove a salivary gland that needed about 14 stitches and drainage tubes.

In common with vitamin C, arnica is good for shock or trauma. Another case involved an unconscious dog whose owner, trying to stop a fight, hit it with a heavy stick. The dog recovered within three minutes of placing the arnica under its tongue. A book on veterinary homeopathy recommends that it is the first mode of treatment in *all* cases, as it calms the patient completely. Available from homeopathic doctors and hopefully vets, as well as health shops, 200 C is the potency most often used for animals.

Arnica montana.

Comfrey

Because it helped heal broken bones, knit-bone was the old folk name for comfrey. It is a broadleaf plant that grows quite readily in damp, cool areas. It will not thrive without plenty of water. Unfortunately, comfrey tends to die back in the winter, but can sometimes be kept going in a sheltered frame where it is protected from frosts. In spite of much publicity to the contrary, it is completely safe both internally and externally. In many parts of

Germany and also Japan, comfrey is used exclusively for dairy cattle fodder during the summer months as it is highly nutritious. Comfrey is also of great assistance when used internally or topically for bone problems, including breaks. It is one of the few plants that contains natural vitamin B12 which may be one of the reasons why it is so good in the case of sickness. Comfrey may be used in poultices and will often reduce bony swellings in a matter of days. It may be made into an ointment or used as a liquid obtained by boiling the leaves. Distilled comfrey oil is the best source of the plant, if obtainable. All forms are useful at some time or other. The plant also has the reputation as an inhibitor of cancer. Like many plants it has a poison (in this case an alkaloid) constituent which if separated from the plant could be dangerous, however fed as a plant it is safe.

The best way to feed comfrey is to offer a few leaves once or twice a week to goats that are stall fed — they appear to find it very palatable.

Comfrey.

Emu Oil

This is now readily obtainable from chemists, fodder stores and breeders as well. It is one of the by-products of emu farming. The oil should be odorless with no additives; it is quickly absorbed through the skin and is very helpful in cases of deep-seated injury. In horses it has reduced bony swellings in the same manner as comfrey.

Emu.

Garlic

This is an onion-like plant that will grow very prolifically if kept damp and well fed. Either the bulbs or the chopped leaves may be given. It is also available in oil-filled capsules or tablet form or in bags already chopped for addition to fodder. It would be suitable for meat or fiber goats, but there would be a risk of tainting milk which would preclude its use in commercial dairy set-ups.

Garlic, like onions, contains natural sulfur and sometimes reduces the incidence of exterior parasites; it is a natural antibiotic, especially useful in intestinal disturbances. Garlic also has the reputation of being a vermifuge and, although it undoubtedly helps, in my experience it cannot entirely take the place of a balanced diet with the correct amounts of copper. In cases of sickness in any stock, persuading them to eat garlic in some form can only be beneficial. It can be blended or offered whole, the farmer must experiment.

Garlic.

Mistletoe

This parasitic plant is a great tonic for goats, well or ill. I pull it down from trees and feed it directly to my animals. Be warned, it turns the urine bright red for the next 24 hours — the goats have not developed bleeding kidneys.

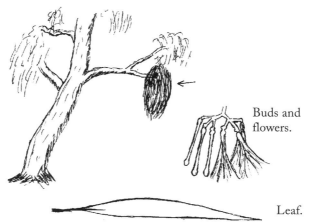

Buds and flowers.

Leaf.

Drooping mistletoe growing on a tree branch (Amyema pendulum).

Parsley

This is a plant high in iron and vitamin A and would be an excellent supplement for cases of acute anemia. According to Juliette de Bairacli Levy it is also very good for edema. I had an aged doe desperately ill with this complaint; she had a badly swollen udder which was pulling her down rapidly (literally). In desperation I looked in Juliette de Bairacli Levy's book and tried the parsley she suggested. The doe had refused all feed and was unable to move, but ate the parsley avidly (I tried it on the other goats — they were not interested). So I gave Catriona a bucketful, which she ate. I had not put any feed out for her and when I came out two hours later to do the milking, she was standing up in a state of high indignation asking for her tea, which she promptly dispatched. The udder was almost normal and I was able to milk her regularly from then on.

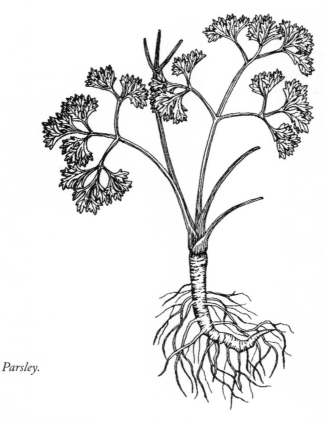

Parsley.

Chapter 11
Health Problems

The first part of this chapter covers CAE (Caprine arthritis encephalitis). The complexity of the disease renders a section on its own necessary. It is only through understanding the implications and testing for this disease that it will be brought under control in our goat populations.

CAE — Caprine Arthritis Encephalitis

This is a disease which is immuno-suppressive and is caused by a retrovirus (sometimes called lentivirus) and, in this particular instance, affects goats. Initially the disease was known as "Big Knees" which was one of the obvious signs when arthritis was present — actually many other parts of the body are affected when the knees are swollen, but that was discovered later. However, we soon learned that arthritis was only one part of it; encephalitis, hard udder, irreparable lung damage (with persistent pneumonia), outbreaks of CLA (Caseous lymphadenitis) that would not clear up, one-sided udders, brain lesions, spinal damage, chronic mastitis and a host of generally unexplainable wasting conditions were all due, directly or indirectly, to CAE. The goat is left without immune defenses against even the most ordinary ailments.

AIDS in humans, Maedi-visna in sheep, Bovine-visna in cattle, equine infectious anemia in horses and so on, every species has its own variety of immune system disease. We can only speculate on the reasons for the sudden upsurge in immune system afflictions. Some schools of thought blame our over enthusiastic use of vaccines, drugs, sprays and artificial fertilizers — our general health has declined seriously and immune systems do not seem to be what they were. All the different autoimmune diseases are remarkably similar, all that differs is the mode of transmission — the net results seem to be the same — a slow, lingering death once the disease becomes active.

There are many different opinions about the frequency and mode of spread in goats (as in other species), lateral or vertical in other words. However, there is absolutely no doubt that the chief mode of infection in goats is via the milk and colostrum. Of course, the blood, as in all kinds of autoimmune diseases, is the greatest carrier; however in properly carried out animal husbandry, infection by blood (via injection needles or tattooing) should not occur.

CAE and Copper

There is a strong link between CAE and lack of copper in the diet. Before anyone knew what it was, the disease had been documented in the United States as a condition where the goat either had not received, or had been unable to assimilate, the correct amount of copper. It seems that a diet deficient in that mineral would predispose an animal to lateral infection.

My goats have always had supplementary copper since before the start of the CAE era (or what we considered to be the start). This was due to Dr. Alan Clark, B.V.Sc. who tested copper levels in my herd so we could establish the dietary amounts needed. Seeing the copper levels are correct would be a small price to pay in the control of this illness. In 1990, in the United States, St. John's wort, a plant high in copper, was first used to help combat AIDS. Also in the United States, the very rapid spread of AIDS had been linked to inadequate copper in the food chain — perhaps due to the advent of plastic plumbing — caught the FDA on the hop. They had always assumed that most people got more than enough copper in their diets and found that when they tested AIDS sufferers in particular, they had only one twentieth of what they should have had (*Acres U.S.A.*). Johne's disease, a simple bacterial condition also needs a copper-deficient host.

I realized that my management of CAE was working and about as bloodless as it could be, but still a nightmare. Others were not so lucky. A certain breeder who was obeying all the rules about separate herds, sheds, etc., rang me to say that yet again her goatlings had come up positive. I asked how much copper she was feeding: "None, everyone told me not to listen to you." I suggested that she take some copper and see if the goats were interested. She took out an enamel pudding dish full of it and 11 goats stood and ate the lot — after that she believed me.

When the copper level tests were done on my herd, I had about one third showing "big knees." Alan Clark and I confidently expected them to show low calcium/magnesium levels as it appeared to be an arthritic condition. To our surprise, they only showed low copper levels, even though I was supplementing with a small amount of the mineral. From these tests, we established the lower level of copper supplementation. I later raised the levels slightly in the diet after reading information from Japan that dark-haired people needed six times more copper than those with fair hair (I ran predominantly black British Alpines). Over the ensuing ten years, while fighting to eradicate CAE from a fairly large herd of dairy goats, I had no lateral spread at all. The only transmission of the disease was by milk and/or colostrum. My goats have a minimum of one teaspoon of copper sulfate a head per week, this is run through the feed on a daily basis as suggested in this book.

Because of having to make my living from milking, I could not afford either to run a double farm (difficult if you are single handed) or wholesale slaughter. I had to do the best

I could, which was to run a mixed herd. This I did for 10 years coming up to 1989 — and probably before that without knowing it. By the time I quit full-time milking in 1992, the herd had been CAE-free for two years.

I have used positive bucks over negative does and vice versa. I have, from the time that I was able to afford testing the whole herd, tried to feed positives together and negatives likewise. Even that came unstuck when someone gave me a negative which was not truly negative; she spent her youth and adolescence feeding with two negatives. In spite of that I have only had one case of a grown animal becoming infected and it was not from the supposedly negative animal. It was doe who managed to raid the bucket into which I put the first squirt of milk from each doe taken off before I start milking. I was called to the phone and, to my horror, when I returned I found she had slipped the chain and had milk all over her face. Two months later she tested positive having had, prior to that time, negative tests and two negative kids. She and another doe that was accidentally infected at birth were the last two before the herd was clear.

Probably the biggest cause of lateral infection is via milking machines. An article from *La Chevre* quoted in the *British Goat Journal* said that it had been discovered that for a few milliseconds, when the clusters are first put on, the pressure in the udder is lower than that in the clusters and the milk is sucked back into the udder. The newest cow clusters in Europe are now being fitted with anti-suck-back shields to prevent the spread of disease. In this country Diversey now markets a small valve that does the same task.

So it is absolutely essential that commercial herds know the status of their goats. Negatives *must* be milked before positives. It is an ongoing program which must be kept up-to-date until the herd is totally free of the disease. Sadly, every country in the world, except apparently South Africa, has CAE. A run down but beautifully set-up commercial concern I saw in Sussex, England in 1988 had reached the stage where 50 percent of the milking herd showed clinical CAE (big knees). Status was totally unknown as was the fact that they had CAE at all. I had the unenviable task of telling them to get the vet in and start testing. The new managers (of two days) knew all was not well, but did not have the slightest idea of the cause. Once they institute a control program, a couple of years should see the situation well on the way to being clarified.

One excellent preparation called VAM (Vitamins, Amino Acids and Minerals), which is an injection available in Australia, enabled me to nurse my positives along so that they could bear their kids. It kept them going and feeling reasonable well when all else failed. In normal goats a two cc intramuscular injection lasts about two months. The CAE goats needed it weekly in some cases, but it was worth it to see them enjoying life.

Testing in the United Kingdom is easy compared with Australia because the test for Maedi-visna in sheep (a disease which is not in Australia yet, thank goodness) can also be used for CAE. The sometimes outrageous prices asked here for CAE testing have been a great setback in bringing the illness under control. Particularly in Victoria where there has been no Department of Agriculture sponsored scheme, as there is in other States.

Catching the kids may be time consuming but, in all cases of straightforward births, it seems to be successful. If the birth has complications and the placenta is broken inside the doe, there is a very real chance that CAE will have been passed to the kid before it is born. I know of two cases where the first one or two kids were "caught" quite successfully, but the last kid was born after much difficulty and it was infected — one was the doe referred to earlier in this section who was supposed to be clear. Kids from such births should be assumed to be positive until they are proven otherwise. Tests must not be done on kids under six months who are fed milk from negative does, and not before twelve months for kids fed sterilized, infected milk. The latter must not be tested until at least four to five months after they have stopped having the sterilized milk. The dead virus will cause passive immunity which will show up as a positive in any test for CAE done before that time.

Kids must be kept separate from positive adults or those of unknown status until they are four or five months at least. Kids that suckle their dams must never be run with positives or goats of unknown status — one cannot be quite sure that they may not suck from the wrong doe.

Care should be taken at shows to make sure the judge's hands are washed between handling each goat's udder. Leaders must open their goat's mouths for the judge and, if you are leading up someone else's goats, make sure you wash before and after doing so.

Tattoo letters and numbers must be disinfected between doing each animal, particularly if they are goats or kids from another farm. Make sure that injection needles are not used on more than one animal at a time — particularly when testing for CAE — regrettably, I once had to tell a vet to use a fresh needle each time when bleeding the goats. These are all possible methods of spread.

In 1991 a further and very disturbing factor emerged, the ELISA test commonly used in detecting CAE picks up a similar signal if the tested goats are sick (with something other than CAE). Only after careful and exhaustive re-running of tests was this fact verified and so saved some perfectly sound goats from death. At last tests on milk are being suggested, the virus was first detected in milk so it is feasible and would perhaps stop any anomalies arising from using the ELISA test.

Catching Kids

Make arrangements for a special kidding area into which all positive does or those of unknown status are removed the moment they show signs of kidding. Have chains with snap hooks at head height on the wall to which the doe can be attached so she cannot reach the kid to lick it. Occasionally kids arrive rather fast. Have clean newspaper ready, catch the kid in it as it is being born and remove it as far away as possible — preferably out of earshot of its mother. Dry and clean the kid with the newspaper, rubbing it quite hard as this helps the circulation, then put the kid away to await its first feed. Some people suggest bathing the kid, I have never done it and all my "snatched" births have been successful.

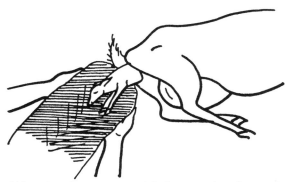

Catching a kid so there is no contact with the ground or the exterior of the doe.

If the kids are taken away in this way, so that the doe cannot see or hear them, it helps stop her fretting. Some people leave unwanted buck kids on positives, but I think it is unwise to take the risk if they are running with negative does as they might suck the wrong doe. One of the sadder aspects of CAE is that the does cannot ever suckle their kids.

Feeding Caught Kids

I have tried feeding kids cow's colostrum, but felt that it was really of doubtful value (and it too, can transmit disease) as immunity is not conferred by any animal save the mother of the kid concerned. Heat treating positive colostrum is a tricky business and it only needs one mistake for the infection to be spread all over again.

I learned to make the first drink from either unpasteurized milk from a clean negative (of at least two generations), or pasteurized milk from positive does, which we all had to do at first. To the warm milk I added one teaspoon of cod liver oil and half a teaspoon of liquid seaweed concentrate (I use Vitec Fish and Kelp Stock drench or the product that Maxicrop put up — these do not have any additives and are the safest). The kids passed their first manure very quickly and never looked back on this regimen and at least there was no chance of some odd disease being contracted from another animal's colostrum.

Heat Treating Milk

This can be done fairly easily by raising the milk to 165 degrees Fahrenheit and keeping it there for five seconds. Use a cooking thermometer and suspend it over the pasteurizing pan so it is in the middle. My first whole season of feeding kids from positive does by this method resulted in all negatives. As mentioned above, do not test until the kids are a year old since before that they could show "passive" positive even though they are *not* infected. Unfortunately, far too many kids were killed until this fact was pointed out by a vet.

Even when the herd has reached negative status, I think it would be very unwise to feed pooled milk to the kids. They should be fed from a select few tested does who are several generations clear. We did, after all, hasten the spread of CAE, which has quite definitely been in Australia for nearly 40 years at least (since 1960 if not before), by feeding pooled milk. In a situation where does always fed their own kids, it could not spread so far or fast.

I remember one quite beautiful black doe that I was given, an excellent milker who showed absolutely no signs of ill health at all, no "big knees" (I would not have known what it was anyway at that time). I always used her milk to feed the kids. When she was nine years old the goats came under great stress from nitrate poisoning and up her knees came. Too late. Most of the kids had been destroyed because we soon learned that any kid whose knees came up — usually at six months or so — became an unthrifty adult, so they were never allowed to live once it happened.

The above story bears out what one of the vets who researched CAE here said to me. He postulated that, in herds where the management was good and there was no stress, he felt that up to 90 percent of the goats could be positives and show no signs until they died and possibly not even then. Many people tell me that they have never seen any signs of the disease so they do not test. But, as soon as those goats are sold to another farm — no matter how good the management — the stress of moving (if they are positive) activates the virus and they start to show big knees, lung troubles, hard udder or whatever.

Never buy a doe unless she is negative, with a vets certificate, unless she comes from a tested herd. A test done while a doe is pregnant is likely to show negative regardless because being in kid often temporarily suppresses the virus in the blood. Does should not be tested until at least two months after they have kidded. When I needed to buy two goats here, as I had lost most of mine in the move, I did not heed the above information. The doe I purchased was from a reputable stud, all advertized as being CAE-free. I had to kill her and her two kids and by that time her milk had infected two more.

I have not gone into details of the disease from the clinical point of view to avoid confusing goat keepers with too much information. Only one fact should be borne in mind; for humane reasons, any goat that shows big knees should immediately be shot or otherwise euthanized. The vet who destroyed six of my positives some years ago emphasized this fact, because the post-mortem examinations showed that the first place to be affected by the virus is the brain (neurotic goats nearly always turned out to be positives), the second sign was spinal lesions and the knees were the last to come up. So by the time the animal's knees showed the effects of the disesase, the goat was already suffering quite badly.

Bucks and CAE

It is obviously important that bucks do not suffer from CAE, therefore they too should be "caught." However, the vets in Western Australia who first isolated the virus and found that it was transmitted in the milk now say that it is *not* passed on by bucks and positive bucks may be used over healthy negative does. This bears out what I have found, I *had* to use positives over negatives and vice versa because I could not afford to do anything else. There was never any transmission of CAE at the time of mating in either direction. This is a merciful dispensation of nature, otherwise we would have lost even more valuable genetic material than we have already. Obviously we want our bucks to last well into double figures, which is what used to happen before CAE, so they should be reared CAE-free.

First Generation Negatives

In my mind there is no doubt that these animals are not quite as robust as the later generations, especially if both the parents were positives. They have to be looked after extra well, after all, it is a small price to pay to be free of what is financially, emotionally and physically a ghastly illness.

General Goat Health

Below is a check list of diseases and the deficiencies that bring them about; I find that they help stock-keepers to realize that diseases are *not* entirely caused by germs.

Goat Check List

Disease	Deficiency
Bone deformities	
Bloat (with potassium)	
Mastitis/high cell count	Calcium and magnesium
Nervous behavior	
Respiratory ailments	
Peeling horns	

Goat Check List *(continued)*

Disease	Deficiency
Tetanies/milk fever	Calcium and magnesium
Warts	
All fungal diseases (i.e., ringworm)	
All worms and fluke	
Anemia	
Autoimmune disease	(spread in absence of copper)
Cancer	
Failing to cycle	
Foot rot and foot scald	Copper
Goat pox	
Herpes infections	
Johnes disease	
Scabby mouth, orf (Herpes)	
Steely fleece ("dermo")	
Failing to hold to service	
Pink eye (blight, conjunctivitis)	Vitamins A and D
Knuckle-over (contracted tendons)	Cod liver oil
Metritis	
Uterine infections	
Lice	
Poor digestion/selenium assimilation	Sulfur
Dystokia	Potassium
Urinary calculi (water belly)	Cider vinegar
Eczema	Zinc
Arthritis	Boron

Unfortunately we cannot be like David Mackenzie and say there should be no health problems since he had the ideal environment for his goats — three miles of coast line and plenty of room. In those circumstances and with the feeding he suggests, illnesses would be at a minimum —

even in the United Kingdom few people can aspire to similar surroundings. In Australia we do not have a hope of emulating him. Here our goats have to face a series of soil deficiencies and imbalances (see Chapter 2) and farmers — of milking goats particularly — have to be very good at their job to keep the goats healthy and productive.

Goats in the wild travel vast distances to find the feeds they want and only suckle their own kids, like their sisters of the meat and fiber sorority, for about three months or less. In these circumstances nutritional stresses should not occur. But those who have brought feral goats into farm situations will tell a very different story, with goats dying by the hundreds in some cases.

The more that we know about the effect of nutritional stress causing disease conditions, the more I feel that hereditary conditions are not quite as frequent as we are led to believe. Several conditions hitherto considered hereditary are now found to be due to nutritional stress at one stage or another.

In some cases the stress is caused by overfeeding; this was much more common about thirty years ago — acetonemia was the most dreaded illness then, but it is hardly ever heard of now. Generally, a lack of the correct minerals in the right amounts is the culprit.

Sick land with calcium, magnesium and/or sulfur deficiencies coupled with a pH so low that the acidity of the soil inhibits the uptake of trace minerals, is also a medium where dung beetles cannot do their task of taking animal manures below the ground. So the result is a double-edged sword, increasing worm burdens suffered by goats who are not receiving the nutrients they need from their grazing. In other words — disaster. In my experience, it takes three years for people to realize the truth of this and then they *have* to take steps to improve their land.

Goats are by nature browsers, worm larvae and eggs are found in wet pasture not up trees, so goats in their natural state would not encounter many worm problems. The same happened in Africa when giraffes were first confined in game reserves and were expected to eat grass instead of the trees they preferred — I'd rather drench a goat than a giraffe.

Nursing

In any illness good nursing and confidence of the patient in the nurse are generally more than half the battle. This is the great argument in favor of well-handled goats. I have bought in poorly handled animals who still, several years later, shrink from human contact (except at milking time); to these animals, drenching or any treatment is an experience so traumatic that often it seems to do more harm than good.

Treatments

Most of the treatments I suggest have been discussed with or emanate from the veterinary profession in many countries. Papers given at various conferences worldwide have yielded much information. Some are learned from an older (and wiser) generation of farmers. For example, the late Mrs. Maura Mackay of Glenroy (Angora and Saanen) fame should have the thanks of countless dairy farmers (both goat and cattle) for teaching us how to use dolomite as a cure, control and preventative for mastitis.

Many people do not notice an animal that is off color until it is really ill — and quite frequently it is often too late by that stage. The goat that is lying down when all the others are standing eating, or is lying apart from the others — that is the animal that should be checked. My method of shedding ensures that I immediately see any goat that does not dive straight into her food night and morning, or any goat that has a dirty back end — both indicators that all is not well. Teeth grinding, yawning and repeated stretching tells the goat farmer that the animal has a pain somewhere and trouble will soon follow if steps are not taken to rectify matters. It is not fair to a vet to allow the goat to reach a semi-moribund state and then expect a cure. Goats often give up when they feel ill and then it will very likely be too late by the time the vet arrives unless some supportive measures like vitamin B12 and vitamin C injections have been used.

I use vitamin C instead of antibiotics for infections, whether bacterial or viral. If it is used in large enough quantities, it works for viruses, unlike antibiotics, and it has no side effects. Antibiotics are used to offset the secondary (generally bacterial) infections which usually occur after a viral infection. Much of the work with vitamin C has been done with friends who are vets and we have been amazed at some of the results when all else has failed.

I have had unfortunate experiences with some antibiotics, but in those early days no one, vets included, knew the right amounts for goats. They were assessed like sheep, as the weights were similar, until several very unlooked for results made us all realize that we had to have a totally different yardstick for goats. All antibiotics have their side effects and I prefer not to use them, although one vet I know of uses vitamins with them and gives good reports of the results. I do not use immunizations — and have found no need to do so — that is my choice. Like David Mackenzie, I see no use for them in properly looked after goats.

It is important to learn how to properly give an injection to a goat. The University of Melbourne taught me years ago *never* to give a goat an injection in the rump or rear of the back legs. This followed a post-mortem on a goat that died from an antibiotic which had been injected into the rear of the legs. I was told by a butcher that animals have a gland around there, and very few people even know of it. The needle had evidently hit it and the leg was already atrophying and the back leg would have eventually become useless. The vets told me that for intramuscular injections, the muscle in the side of the neck was always absolutely safe and *never* to give any injection without thoroughly cleaning the site first.

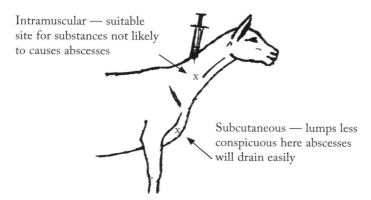

Always inject a goat intramuscularly in the side of the neck, a site that is fully safe.

Intravenous injections are often very good if there are two people present, one to hold the animal still and one to inject. But if the goat is in a state of shock, or when the veins collapse, it is no use trying to find one. This frequently happens in the case of snake bite.

The bottom line in any sickness is good nursing, keeping a goat warm and as happy as possible under the circumstances. This can pose problems because sometimes a goat will fret if removed from its companions. A sick bay within sight and sound of the other goats is sometimes a good idea — other times they are better totally segregated — one has to play it by ear. It is no good just giving the animal the appropriate treatment and leaving it to sort itself out, they need care and reassurance.

Abscess

An abscess can be caused by grass seed working out from the back of the mouth or an organism entering a break in the skin. The body tries to expel the foreign matter and an abscess forms in the process. It can be hurried up by hot fomentations, or occasionally extra vitamin C, but it is really best to leave it to come to a head naturally. The vitamin C can be given as an injection of three grams for two days running which speeds up the process by detoxifying the poison.

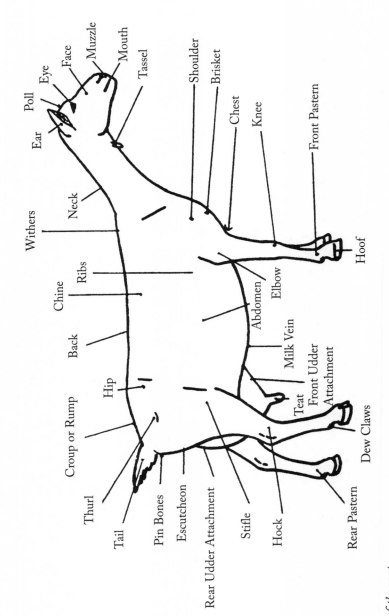

Parts of the goat.

When the abscess comes to a head and breaks, clean the pus away — burning all material used in the process. Wear rubber gloves if you have broken skin on your hands. Then syringe out the site with a mixture of 10 percent copper sulfate and water, trying to remove the core of the abscess. Allow it to drain and close up of its own accord and be careful it does not become fly blown. Flints oils or septicide ointment are both good preventatives against fly strike and the wound can be filled with either.

If an abscess is allowed to come to a head naturally and cleaned out as described, the healing will be very fast indeed. If however, the abscess is lanced before it is ready, a very nasty mess ensues which often takes several weeks to clear up.

Grass seed abscess.

Acetonemia

This used to be *the* goat disease when I started goat keeping at the beginning of the 1960s. It was almost endemic in studs where goats were fed a great deal of high protein feed without the balancing carbohydrates and dolomite. This was done to encourage high production and many of the goats were over fat and under exercised.

Disease signs are misery, irregular cudding, lack of appetite, dark, sticky looking droppings, and breath smelling of pear drops. Remove the cause and treat first as a cobalt deficiency, giving two cc of vitamin B12 by injection three days running and a dessertspoon (9 gr) of dolomite daily for three days. Rethink the feed program and see that the goats are exercised.

Goats that are heavily hand-fed become lazy about going out to look for grazing, a bit of mild starvation will usually give them the idea. Goats that regularly receive the correct minerals and whose food does not exceed 14 percent protein should not succumb to acetonemia.

Anemia

This is due to a shortage of hemoglobin, or red blood cells. In any country except Australia this could mean a lack of iron in the fodder. In Australia most soils have adequate or too much iron so it will mean that the anemia will be due to a lack of copper. Without copper, iron cannot be utilized (see section on copper). The other causes could be liver fluke, blood-sucking worms like barber's pole (*Haemonchus contortus*), brown stomach worms (*Ostertagia*) or bleeding from an internal injury.

The most obvious sign of anemia are goats that are lethargic and off their feed. Examine the membranes of the eyes, they should be a good deep pink to red but will possibly be a rather pale pink to white. Checking the membranes of the eye should be a weekly, if not daily, part of good husbandry.

If lack of copper is the cause of the anemia, this can be fairly easily adjusted if the animals are fed copper through their ration as suggested in Chapter 6. But a worm count should also be taken because blood-sucking worms kill goats, especially kids, very quickly indeed. Kids with barber's pole infestation will be found to have chalk white eye membranes. Act *very* fast; administer vitamin B12 injections (one cc) every four hours, the mildest possible worm drench (as a very strong drug could kill at this stage), and for the next three days provide an iron tonic (Ironcyclene or similar). Give them seaweed meal ad lib which will go some way to building up the copper, which should be in the feed as already described. Worms and fluke are not interested in hosts whose copper requirements are being met.

Uncharacteristically low butterfats are often due to anemia, they are not always hereditary. A failed milk test is quite often a goat owner's first clue that something is wrong. I was told years ago always to give the goats a course of vitamin B12 injections coming up to a milk test (where butterfats are recorded). Sometimes I did it, not knowing why and wondered if it made any difference, but obviously the person who had told me found that it did.

Arthritis

Beginning from the late 1970s until now, we have come to equate this condition with CAE. However, goats can and do get arthritis that does not owe its origin to that virus.

Signs are creaking joints (audible a few feet away) in mild cases, heat, stiffness and sometimes swelling in the joints — knee and stifle in particular. Arthritis is caused by nutritional stress due to an imbalance of the minerals in the feed. When it occurs in animals that are already receiving the correct amount of dolomite, it will be due to insufficient vitamins A and D and/or copper and/or boron — all are needed to assimilate calcium and magnesium.

One of the papers presented at the International Conference on Goats in Tours also implicated a lack of copper as a predisposing factor (copper bracelets on horses have had good press). The lack of vitamin A and D can be due to reasons suggested in the section on those two vitamins (see Chapter 10). Lack of boron necessary for vitamin A and D absorption will be due to a shortfall in the soil. One teaspoon of borax between twenty goats once a week supplies enough boron.

Treat arthritis by removing the grains from the ration initially — give plenty of good quality grass hay, green feed and a little chaff and bran. Give vitamin A and D in some form regularly and include seaweed meal ad lib for the boron and copper. Cider vinegar is also a great help and should be added to the feed, or let the patient help itself. As much borax as will adhere to the tip of a finger can be fed daily. These days, in any case of arthritis, the goat must be tested for CAE and if it is positive no treatment will work.

Arthritis — Infective/Septic

This is caused by an organism that has gained entry through a wound, or possibly the navel cord (which may not have been properly disinfected at birth), or, more rarely, following mating to an infected animal — this can work either way. It usually takes about six weeks for the organism — generally Corynebacteria — to show up. It happens very suddenly, with arthritic symptoms and a high temperature. Very occasionally high doses of antibiotics work, but this bacteria is notoriously difficult to treat, especially when it is in a joint where it can cause irremediable damage. Vitamin C therapy started immediately when signs are observed could possibly be successful — give a kid five grams (10 cc) intravenously if possible, if not, inject in the muscle and repeat every 12 hours. Give half a teaspoon of cod liver oil every two days, preferably by mouth. Good nursing procedures and giving afflicted adults twice the above dosage may work. Make sure the patients have unlimited access to seaweed meal.

Infective arthritis, whether from the navel or otherwise, is an unfortunate condition because it lies dormant for quite a long period while the causative organism is already doing damage. I bought a kid from interstate which was seven weeks old before a navel abscess showed up closely followed by infectious arthritis. We did not know about vitamin C in those days and the available drugs had little if any effect. It is also probable that it would have been too late for vitamin C to work. Its navel cord had not been disinfected and it was born in an old sheep yard, a frequent source of infection.

I learned about the venereal variety when a doe, who I afterwards learned had aborted previously and was in poor health, was brought to a good young buck for service. The first I knew was several weeks later when the buck suddenly became crippled with arthritis in all four legs.

Everything was tried, and eventually he was put down and a post-mortem was performed. The cause was then discovered, Corynebacteria and a swab of the last four does he had covered revealed the culprit, who appeared perfectly well. She did not hold to the service and was probably incapable of conceiving.

Do not let a doe come to your buck if there is any history of abortion or similar trouble. Insist on a clean swab (a swab can only be done when she is in season) before she comes back.

Avitaminosis

This condition literally means that the goat has run out of essential minerals and therefore vitamins rather suddenly. Unusual lethargy, unwillingness to move, eat or drink are the first signs of this ailment. Examine the membranes of the mouth, according to the severity of the condition — they will either be streaked with scarlet lines or be a bright pillarbox red all over. Give the affected goat a dessertspoon (2.4 teaspoons) of dolomite, the same of vitamin C, and two ml of VAM in the muscle — also give seaweed meal straight into the mouth and leave it for the animal to take as much as it wants. Usually this is enough, but in a severe case the treatment may be repeated eight hours later. I first met this affliction in a doe belonging to a friend who had bought the minerals for it but had not fed them. I suggested that she weigh the minerals, put them out for the goat and see what happened — the goat in question ate a pound without stopping and recovered almost instantaneously.

Bent Leg (Kids)

This is caused by a bad calcium/magnesium to phosphate imbalance which is almost invariably produced by overfeeding with milk. In one case where I was consulted, a kid was being given 12 pints of milk a day — far too much. The bones (and the kid) grow very fast when this happens, too fast for the bone to form as it should — hard and flinty. The imbalance makes bones soft and porous, so the weight of the body causes the legs, usually front, to bend.

To treat it reduce, preferably stop, the milk intake, (see Chapter 8 on the reasons against overfeeding kids with milk) to about a pint a day at most. Give a teaspoon of cider vinegar daily and some form of cod liver oil, (vitamin A and D injection or by mouth). Grain must not be fed, only a little oaten chaff, bran and good grass hay. Dolomite must be mixed in the feed, a teaspoon a day, and the same for yellow sulfur — no more. Seaweed meal should be obtainable ad lib, and a gram of copper a day in the feed (dissolved in the cider vinegar). If caught in time, the legs usually straighten by the time the kid is full grown.

Bent Leg in the Newborn (Contracted Tendons)

In this situation the kids are born with their front legs bent under them so they stand on their toes — this can affect either one leg or both legs. The impression is that the flexor tendons are pulled up too tight, which is exactly what has happened. In extreme cases the leg or legs assume an "S" bend. I used to think this was just one of those things that would right itself — it usually did — eventually. Some authorities list it as being hereditary, but it is not. Drought or overhead main grid power lines can both interfere with the correct synthesis of vitamins A and D.

My last untested (for CAE) doe to kid produced two does with badly contracted tendons. I gave them the homemade colostrum mentioned in the section on CAE. Within a few hours, the kid's legs had straightened completely instead of taking the usual week. The next doe to kid was from the same place as the first one; both had ongoing vitamin A deficiencies, possibly due to the fact they were reared under power lines. Again the kid's front legs were deformed and the vitamin A and D worked just as fast. In this case the mother was able to feed her own kids, so I gave each one half a teaspoon of cod liver oil straight in the mouth.

Recently a friend called me with a six-week-old kid that had a contracted tendon in one leg only. It was so bad that the leg was twisted right around. I felt that in one leg there could be a deformity, but suggested that she try the vitamin A and D. To her amazement (and mine) the kid came good in a few hours.

Goat farmers should make a note of does who produce kids with this complaint and give them a cod liver oil booster about three weeks before kidding. This should prevent contracted tendons from occurring. The does could be given either vitamin A, D and E injections, or oral cod liver oil.

Bent Leg in Adults

This is usually caused by a sudden increase in phosphate-rich feed resulting in a possible withdrawal of calcium and magnesium from the bones. The goat pictured

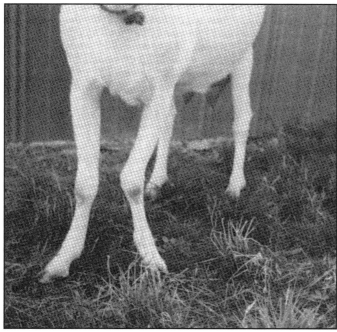

An adult doe with bent leg due to overfeeding of alfalfa.

here was given extra alfalfa hay coming up to kidding because it was a particularly hard winter. The diet should have been altered by giving extra oaten chaff, bran and little barley with as much good quality grass hay as desired. I also recommend the following additions to routine supplements:

> Give 1 tablespoon of dolomite with each feed for a week, then cut it back to the regular dessertspoon (2.4 teaspoons)
> 1 tablespoon of cod liver oil
> 1 pinch of borax in the feed twice a week
> 1 cup of cider vinegar with each feed
> Seaweed meal ad lib

Once the leg starts to harden up, go back to the regular ration but avoid sudden increases in rich feed.

Beta-mannosidosis

This is an hereditary cell storage disease, similar to Alpha-mannosidosis in Angus cattle. The cells cease to function properly and store material that should be expelled until finally the system, loaded with toxins, breaks down. The brain cells appear to be the worst affected. Up until the present time in Australia this problem has mainly been found in Nubians and a few Saanens. The signs are usually fairly conclusive and it shows up in kids as spasticity. There appears to be no known cure and affected animals usually live no more than a few weeks.

Owen Dawson visited one property where Nubian bucks had been used over ferals for meat production and there was a very high proportion of spastic kids. Thus it would seem that it possibly is not necessary for both individuals to carry the disease — perhaps it can be transmitted by one animal. Certainly in one Nubian stud that I knew well — from a herd where the status of all the animals was known — the transmission was very uneven, often two positives did not produce a spastic kid, then, or in later generations.

Beta-mannosidosis appears to be another of those inconclusive conditions. It is thought to be hereditary and to need two carriers. But considered opinion in many cases links this condition to management. In studs where goats are well looked after it has rarely been a problem; lack of viability due to inbreeding may have triggered it off originally.

Blackleg — *Clostridium chauvoei (Chauvoei, Bacillus chauvoei* and *B. anthracis symptomatis)*

This clostridial disease is caused by a scratch or surface wound, often on old sheep country, that has not been disinfected properly — again because it may have been too small to see. The first lesson the vets I worked with in the United Kingdom hammered home was that a wound must be properly disinfected at once. The vets used peroxide, iodine, disinfectant (Lysol in those days) and said that if all else fails alcohol — gin, whiskey or whatever (a 10 percent copper sulfate solution is also very effective). This disinfectant must be syringed into the wound if necessary, and all dirt removed if possible.

Blackleg is rare in goats in Australia and is reported in *Hungerford's Diseases of Livestock* to be incurable. As of 1990 only three cases had been reported here, two indirectly due to five-in-one vaccinations. In the first case, a dirty needle was used, and the other followed two days after a routine five-in-one vaccination where every care had been taken. The vet concerned in the latter case said that in the future he would not recommend this type of vaccination for goats, but rather would use only the two-in-one (tetanus and enterotoxemia).

The third case followed a goatling being cut along the side by barbed wire on an old sheep farm. The wound was not cleaned and three days later the first signs of blackleg appeared. The goatling was saved, but it took three weeks because, when the owner first rang, I did not realize there had been a wound. So I imagined it to be a bite and not blackleg. I advised a small injection of vitamin C, five grams, which kept it alive but meant the cure took two weeks. I now know that the treatment was inadequate and the suggestions below work in hours not days. This goat was the fourth case I knew of and happened on my own farm. I felt the goatling had pierced her leg just above the stifle joint, but there was nothing visible.

In blackleg the limb, generally a rear one, swells to grotesque proportions and the goat is in great pain, usually lying with the leg sticking straight up in the air due to the swelling. If no action is taken, shortly afterwards the head starts to swell and the goat will die very soon from the enormous pressure of the swollen parts, which rupture and turn black, giving the disease its name.

Do not try injecting into the neck as usual, because blackleg makes the whole body super sensitive. Inject straight into the affected limb repeatedly (every few hours) with 25 grams of vitamin C, give good supportive nursing and an injection of VAM and vitamin B12 in the neck. As soon as the goat is eating again provide ascorbic acid powder in the feed. Continue to inject the affected limb with vitamin C until it goes down (about 24 hours). Find and disinfect the cause if possible.

The goatling on my farm, seen grazing in the garden four days after the blackleg attack.

Bloat

This condition is caused by potassium and magnesium being unavailable — generally in an overly rich pasture where clover is dominant. On a minerally balanced farm, the clover and grass are equal and bloat does not arise, however good the year. Tallow is another cause of bloat in kids which have been fed a milk replacer that contains processed or just plain tallow. Both stop the kids from obtaining the necessary nutrients from the milk and they die of bloat (and starvation). No therapeutic measures work when bloat is caused this way. This is because the tallow coats the inside of the alimentary canal and no nutrients can be absorbed.

In bloat the goat's abdomen will be much distended, especially on the left side. If the goat is still able to walk, drench a quarter of a pint of cooking oil down the throat, then exercise while massaging the sides. This usually persuades the wind to be passed from one end or the other. As soon as the goat is relieved, give a dessertspoon (2.4 teaspoons) of dolomite mixed in half a pint of cider vinegar which will help replace the missing magnesium and potassium.

If the animal is down and in distress, call a vet immediately because the pressure in the abdomen will quite soon stop the lungs and heart from working. The vet will release the gas with a trochar (a sharp, hollow surgical instrument with a retractable center) allowing the gas to escape. The incision is made four fingers width behind the bottom of the ribs on the left side of the goat as it lies.

If the vet is unobtainable, a sharp, pointed knife will work in an emergency. Disinfect first, insert the knife point until the gas starts to escape, twist it slightly, remove the knife, and close the wound once the distension is relieved.

Again, a drench of seaweed meal with dolomite and cider vinegar (about 10 fluid ounces altogether) should be given as soon as possible to build up the magnesium and potassium in the system.

With bloat, prevention is easier than the cure — or death. Have the paddocks analyzed as soon as possible and top dress with the necessary lime minerals. Do not use chemical fertilizers under any circumstances. If the bloated goats are still feeding on the paddock that caused the trouble, make sure that all the necessary minerals are in the ration. Seaweed meal should be ad lib as usual.

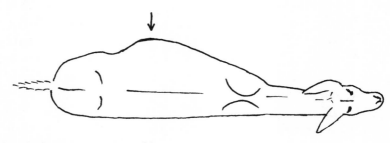

Site for letting gas escape in bloat.

Blood in the Milk

Unless caused by a blow from another (usually horned) goat or animal, blood in the milk usually occurs in first lactation milkers when the udder is expanding rather fast. It is the result of ruptured blood vessels in the udder. If accompanied by viscous or offensive-smelling milk, treat as for mastitis (see mastitis later in this chapter).

Treatment is to give at least a teaspoon (five grams) of vitamin C powder orally and six cc by injection. Repeat the oral dose daily until the milk is clear. Milk with blood in it must, of course, not be included in a contract. To check that the milk is clear, leave it in a glass container for 24 hours — the blood will be seen as a dark line through the bottom of the glass.

Boils

See Abscess.

Bottle Jaw

This is a hardened swelling under the jaw and is almost invariably caused by worms or liver fluke — occasionally it will also arise in cases of extreme debility. Dealing with the worm problem usually clears it up, consult the section on worms. If this is not the cause, attend to the goat's general health. If the goat's CAE status is unknown, have a blood test taken. If the goat is positive it can only become worse.

Broken Bones

An ordinary bone break where the skin is unbroken is fairly easy to heal in a light-weight animal like a goat. If the fracture is compound (where the bone protrudes), call the vet quickly.

When splinting, pull the limb out carefully so that the ends of the bones meet and try to set it in that position. Plaster of Paris is not really a success on small animals, as it is difficult to keep dry and tends to be too heavy. A leg can easily be splinted — first apply a soft bandage the full length of the affected limb. It is usually better to bandage the whole leg, especially if the injury is above the knee or hock. Then put on the splint. Splint materials include the side cut out of a plastic bottle, a metal "ladder" (obtainable from a vet), split bamboo or other light wood, which all work equally well. Then bandage the splint in place and sew the end of the bandage. Do not use safety pins, clips or knots, as they can end up inside the goat or be undone (or both). Bandaging is an art — too loose and it comes off or too tight and the limb drops off. Experience teaches you plenty, so check an hour later to see that pressure has not built up because the bandage is too tight.

Years ago I treated a two-month-old kid for a broken bone who was staying on my farm and she was returned, unblemished, to her owner. Standing beside the animal at a show years later, I told the owner to feel carefully down the lower part of the leg; she could just feel a very faint ridge where healing had taken place.

Leg with splint and bandage.

Goats are excellent patients; they never attempt to put weight on an injured limb until the fracture is healed. It is best to keep them away from long undergrowth while the splint is on, otherwise the leg may become caught up and twisted. Do not keep the patient indoors, it is better off leading a normal life. A healthy goat will be fully healed in ten days or less. At that time remove the splint, but leave the leg strapped up for another week as a support.

In both cases, compound and ordinary fractures, add one 500 IU capsule of vitamin E and three to five grams of oral vitamin C to the diet, plus one teaspoon of cod liver oil a week. An extra teaspoon of dolomite daily should also be given. Those who have comfrey growing should feed three to four leaves a day to a kid and double or more to an adult goat. If you can find comfrey tablets, give two a day for a kid, and four for an adult until healing is complete — goats will often chew the comfrey tabs up — otherwise crush them in the feed.

Brown Stomach Worms

See worms.

Cancer

This illness seems to be on the increase in all animals, probably for many of the same reasons that it is increasing in humans. Goats, like most animals, have a head start over us because they manufacture their own vitamin C and are therefore never without it in the system (see section on vitamin C in Chapter 10). Fluoridated drinking water can be a contributory factor in cancer (see section on magnesium in Chapter 10).

The following treatment has been used successfully on a number of different animals and would be worth trying. For an adult, give five grams of vitamin C, with one cc of vitamin B12 (in the same syringe) daily for two days. Then five grams of vitamin C alone by injection for the next week. Add a dessertspoon (2.4 teaspoons) of vitamin C daily in the feed as well and continue the oral dose for another month if the cancer has gone down.

In addition, 50,000 units of vitamin A and 1,000 units of vitamin E must be given daily for two or three weeks, then bring the dose down to 10,000 units of vitamin A and the same vitamin E as before for another two weeks (dissolving the capsules is the easiest).

Feed the normal minerals as usual, including the copper (which helps the immune system) and make sure seaweed meal is available on demand. Andre Voisin links cancer with lack of copper.

If no improvement at all is noted in ten days, it is probably kinder to put the goat out of its misery. Actual tumors should regress totally in six weeks or less, but a large improvement in general well being, lessening of pain and decrease in the size of the tumor should be seen in about ten days or less. If so, continue with the vitamin A, E and oral C daily until healing is complete.

Tumors rarely disappear altogether and most often a small node is left — keep an eye on it, if it starts to enlarge institute treatment once more. Occasionally if the tumor is on the surface, it will discharge like an abscess after a few weeks. This should be cleaned up as suggested in the section on abscesses. This is preferable to the tumor dissolving inside where it inevitably sets up toxicity — hence the elevated doses of vitamin C to prevent any reaction.

During treatment the goat should be on a grain-free diet with really good plain grass hay and grazing. If the goat contracted the cancer on your farm, analyze and top dress the paddocks with the lime minerals as fast as possible.

Cane Toad Poisoning

Give vitamin C by injection and orally as for snake bite.

Car Sickness

See travel tetany.

Casein in the Milk

This is a fairly unusual complaint. I have only met one case, but I understand some does are prone to it. Obviously something is wrong with the metabolism, but the cause is unknown. I never had another case once I started to feed dolomite and cider vinegar regularly, so it is possibly dietary.

A hard lump forms in the teat, which is almost crystalline in appearance when extracted. Massage it *very* gently down the teat and try to persuade it through the orifice, causing as little pain and damage as possible. Make sure the teat and hands are spotless, and give a teaspoon of dolomite with the same of vitamin C powder each time to counter the risk of mastitis.

Caseous Lymphadenitis (CLA), Cheesy Gland

This is quite different from a grass seed abscess, although it may take a vet to tell the difference. The latter has been covered at the beginning of this section. CLA is due to an organism — *Corynebacterium pseudotuberculosis* — which gains entry through a wound, often invisible, or even from a grass seed. The abscesses are located on the lymph system and usually start at the back of the jaw. They then follow the lymph system down via the shoulder and underarm to the stifle from whence they will form inside the animal, usually resulting in debility and death. If the goat's immune system is in good order, one abscess is usually the only result, but should the goat be CAE positive, with no natural immunity, the abscesses will, in my experience, become endemic.

The abscess starts as a flat hardening at the back of the jaw, developing into a boil varying from the size of a dime to that of a tennis ball, depending on the goat's natural immunity. The treatment is the same as described earlier in the section on abscesses, *but* extra care must be taken in the handling of the pus from the abscess. Rubber gloves *must* be worn if there is any likelihood of a skin break in the hands and all material from the cleaning must be burnt. A British Veterinary Society meeting in April 1990 classifies CLA as a zoonoses, in other words, an animal disease that can be contracted by man, often in a form more unpleasant than the original.

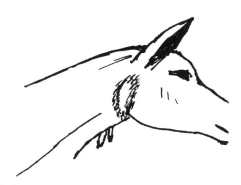

Cheesy gland abscess.

I had an experience with an outbreak of this disease many years before CAE became a problem. Every goat in the herd, about 15, produced an abscess of some kind (often very small). From then on, even if a new goat came with an abscess, my herd appeared to have built up a natural immunity that lasted for years. The organism re-sponds to no known antibiotic; exhaustive tests were done at the time of that outbreak by veterinary researchers. They established that it could sometimes be halted if measures were taken before the swelling started — an obvious impossibility.

The only measures I have taken have involved using megadoses of vitamin C when the abscesses are internal. This does not stop them, but it detoxifies the material from the abscess so it does not affect the system — this, of course, only applies to CAE-free animals.

Many years ago a woman rang me about a goat in the last stages of debility. About three boils had burst on the exterior of her doe and then started through the interior lymph system. The vet had diagnosed them as being in the lungs and liver and advised putting her down and I felt the same. The owner believed that while there was life there was hope, so we decided to try vitamin C. The doe was given 10 grams intramuscularly daily for 10 days, with several injections of two cc of vitamin B12 and good supportive nursing. I did not hear any more and assumed the doe had succumbed until several months later when I opened a letter and a picture of an absolutely blooming Saanen doe fell out — she had made a full recovery.

In some countries, the United Kingdom included, CLA is a notifiable disease. It most often emanates from sheep, among which it causes havoc because the abscesses cannot be dealt with or seen in the wool. Exporters of goats may have to produce certificates of freedom from the disease.

Several attempts, mostly in South Africa where it has been a scourge to fleece goat breeders, have been made to produce a vaccine. These apparently have not been very successful. It was found that good husbandry, such as avoiding deep litter situations and making sure that the kidding yards were never on the same ground two years running, were found to be far more effective.

Coccidiosis

This is caused by an order of parasitic protozoa. There are several varieties and each is species specific. In other words, each animal has its own particular strain. Goats cannot catch it off birds or cats, but birds can carry it from another of the same species. I personally have doubts on that one since cases of this illness have arisen — a specific instance was in a horse (which is rather rare) in the United Kingdom and no other horses were known in the county with the disease.

Apparently many older goats can carry, and shed, the disease all their lives, showing no signs of actually having it although a worm count might reveal it. For this reason, especially in small goat operations, kids are better segregated from the older goats for grazing. All animals develop an immunity with age.

Signs are ill thrift that does not respond to the usual measures and, in severe cases of coccidiosis, blood will be seen in the droppings. A test of the manure would confirm its presence. The drugs used to treat the condition are severe and prevention is best. Rudducks used to make an oral or injectable mixture of sulfadimidine for treatment that was the least traumatic of those on offer.

I prefer to make sure that all goats, especially kids, have access to mallow (Malva) plants. This herb has the reputation of preventing coccidiosis. Lately I have seen it listed as poisonous — it is no more so than any other herb when eaten in excess. Certainly I never had any more problems once I made sure it was growing round the kid yards.

Also it seems that properly supplemented stock of all kinds that are receiving the right amount of copper do not contract coccidiosis. This has been especially noticeable in the dairy cattle industry where it was a scourge in calves. Owners report that since proper feeding of minerals was started they have had no more cases.

Congenital Defects

See hereditary defects.

Contracted Tendons

See bent leg.

Coughing

A vet friend of mine used to be able to make a goat cough by exerting slight pressure under the neck. It will be noticed that goats often cough while being led if they have not been properly trained and are pulling.

Persistent coughing with no sign of fever can mean lungworm (see worms) and any goat that has had lungworms is often left with scarred lungs and will cough intermittently for the rest of its life. A course of 500 units of vitamin E daily for one or two weeks could help clear it up as this vitamin minimizes scarring.

If the coughing is accompanied by distress and a high temperature, treat as pneumonia — start treatment as quickly as possible. See section on pneumonia.

Respiratory problems traditionally affect goats or any other stock whose calcium/magnesium levels are not correct. Have the paddock analyzed and do the necessary remedial work.

Chilled Animals

Adults

These should be brought into a sheltered area and friction applied. Tie a jute sack around them if possible. Hay must be offered ad lib, food in the rumen is the goats natural way of keeping warm. If the animal is hungry, cold stress can occur. Give a heaped teaspoon of vitamin C orally and repeat two hours later, 10 cc injections can be given if preferred. See pneumonia section for further treatments if necessary.

In spite of warnings to the contrary, a big teaspoon of brandy in some milk — an old fashioned remedy — is still highly effective on occasion. To heat up a chilled large goat, pour methylated spirit along the spine and immediately rug it up and keep the goat warm. This remedy has been used effectively to warm up animals — it depends on the latent heat of evaporation.

Kids

These can become chilled right through and will require a lot of attention if they are to live. Put a finger in its mouth and if it feels cold all way down, bring it inside by a fire (the bottom oven of a slow combustion stove, with the door open, is good). Otherwise keep it wrapped up. Give it a little warm milk with half a teaspoon of brandy in it. I know this method is frowned on these days, but I have saved countless lambs and several kids using it.

I had a kid which fell into the irrigation channel in the winter. I did not find out until feeding time that she was missing. She had possibly been in the water for four hours and she was cold right through. I gave her a teaspoon of brandy in milk, six cc of vitamin C by injection and wrapped her up and placed her by the bottom oven. She was still terribly cold when I went to bed, so I repeated the treatment and took her, well wrapped up, along to bed. I wanted to know how long it would be before she warmed up. It was two a.m. when it happened and she finally relaxed and slept. She suffered no ill effects.

Circling Disease

This can be a result of listeriosis, occasionally Corynebacteria or a nasal bot fly larvae that has got into the wrong place, affecting the brain. All these can make the goat one-sided and the animal then circles incessantly. The gait is stiff and the animal walks with its head stuck forward; post-mortem will confirm which was the cause. If it was a bot, the meat will be quite safe to eat, not otherwise. In any case, the ailment is incurable.

Cow Hocks

These cannot be classed as an ailment. Though for many years they were considered hereditary — and probably were in the United Kingdom originally where deficiencies were not so frequent as here.

In Australia all too often goats with perfectly straight back legs produce offspring that become cow hocked with age. I have judged young kids with really good back legs, only to meet them again a year later with their hocks almost touching. A course of cod liver oil, (orally, as the straight liquid, an emulsion, or as an intramuscular vitamin A, D and E injection) should be given. Assuming, of course, that the goat is receiving the correct minerals in its feed and has seaweed ad lib available (see Chapter 6). Very young kids whose back legs suddenly go weak can be given two or three cod liver oil capsules (as for humans), which will effectively reverse the condition.

Cow Pox (Goat Pox)

This is a herpes linked illness. Small pustules appear on the udder and sometimes around the tail and mouth. If unchecked it can spread into large sore scabby areas. I have seen a buck who was literally covered with pox and had to be treated with large doses of vitamin C to arrest secondary infections. Like many herpes diseases, goat pox is supposed to run a three week course. In my experience, if goats are copper deficient it can become almost endemic.

To treat the exterior, make up a wash of a tablespoon of copper sulfate and the same of vinegar in about a pint of water. This can either be administered from a garden spray bottle or rubbed on with a sponge, the scabs will dry up and start to drop off. However, the disease is more prevalent in colored and black goats whose copper requirements are not being met. A British Alpine goat that I saw at a show, which was slightly afflicted, was given half a teaspoon of copper sulfate in her feed for two nights in a row and the condition cleared up without any exterior treatment.

First lactation does who, because they had not been receiving the same amount of copper as the adults due to lighter feeding when goatlings, always seem to be prone to pox.

Goat pox is occasionally infectious to humans. Most people who milk by hand will have built up an immunity, especially if they have had chicken pox.

Cystic Ovaries

See hereditary defects.

Dandruff, Scurf

This is caused by a deficiency — or excess — of iodine. In goats that have been on the farm and have either been on the stock lick (if fleece or meat) or ad lib seaweed meal (if milkers) this should not arise. However, should goats arrive from another farm with bad dandruff, check that the previous owner has not been using too much iodine in some form. If seaweed is fed ad lib, goats with an excess would not touch it, and if caused by a deficiency, they will make it up from the ad lib meal.

Dandruff can be a problem in fleece goats as it can affect the quality of the clip, but more seriously, it means any goat that has it is below par — read the section on iodine. In dairy animals dandruff should be taken seriously as a sign of an iodine deficiency, as well as being a nuisance if the goats are shown. Make sure that the stock lick is out for the meat and fiber goats and ad lib seaweed meal for the milk goats.

Deformities

Some of these are hereditary and are listed in that section. If a kid is born with a deformity (and not contracted tendons, see that section), it is probably hereditary; but if the condition develops later it is more likely environmental (nutritional). Lack of magnesium can and does lead to postnatal bone abnormalities, but this can occasionally be caused by overfeeding of protein.

One of the more usual and apparently hereditary deformities are twisted and either under or overshot jaws and, of course, abnormal teats.

Some years ago I sold a perfectly normal kid which, as usual with stud animals, I had photographed first. A few years later I heard that it had developed a parrot mouth by the time it was two years old. I could not understand what had happened until later on when another goat was brought back at two years of age for mating; so bad was the deformity of the mouth that, until I checked the tattoo, I doubted if it was the same goat. The doe duly kidded and came back to be mated; this time I really thought I was seeing things because the mouth was 100 percent normal. I asked the owner what she had done to accomplish the miracle.

"Oh, I knew you were always rattling on about dolomite, so I started feeding it and the mouth gradually became normal in couple of months."

If the damage is allowed to persist into old age, it may be incurable. The cases I have seen in goats had been cured before the animals reached full maturity — four years for a goat — but I know of a 14-year-old horse whose bones normalized after being fed correctly for a year, so perhaps there is no time limit.

Dermatitis

This is rather like goat pox in appearance. Symptoms are pustules on the udder, which can also spread to other parts of the body as the goat rubs the udder with her mouth, then scratches her face with a foot, etc. Unlike goat pox, dermatitis seems to be fairly contagious. Contaminated teat dip cups, udder cloths and dirty hands can all spread it. A copper wash made up of a tablespoon of copper sulfate, the same of cider vinegar in about a pint of water and used as a spray will help clear it up. A vitamin A, D and E supplement should also be given along with half a teaspoon of copper sulfate with a teaspoon of ascorbic acid in the mouth for two consecutive days. The normal dose of copper in the ration should be given along with the above as well. After two days the vitamin C alone may be given to help recovery and avert secondary infections from the lesions.

The condition is staph related and likes a copper deficient host, but is not serious enough to warrant using a vaccine as is often suggested. Goats appear to develop a natural immunity after a while. It is noticeable that in herds with a high incidence of the complaint the dietary copper is nearly always non-existent.

Diarrhea (Scouring)

Intestinal worms and infections, cobalt, copper-deficiencies and enterotoxemia are the most usual causes of this complaint. Other reasons can be imbalance in the feed, paddocks too high in nitrate-rich feed such as capeweed (already mentioned), and feed that produces acidity. In small kids overfeeding of milk often causes mild scouring.

For an adult a dessertspoon (2.4 teaspoons) of dolomite, a quarter of a teaspoon of copper sulfate and the dessertspoon (2.4 teaspoons) of vitamin C down the throat is always worth trying first, especially if there are no other signs of illness. If the goat is listless, lacking in appetite or has cold ears, suspect a cobalt shortfall and give two ml of VAM and two ml of vitamin B12 intramuscularly. Quite often both or either will clear the scouring up. Scouring is not (as is generally supposed) always caused by worms. *Hungerford's Diseases of Livestock* states that unexplained scouring is often due to a copper deficiency and sheep farmers have found that weaner lambs have responded to half a teaspoon of copper and the same of dolomite quite remarkably when all else failed. The same could be tried with young goats, the dolomite should always be given at the same time (a teaspoon of dolomite and a quarter teaspoon of copper sulfate).

If intestinal infection is suspected, treat as for enterotoxemia (see section on that ailment).

Scouring, especially in kids, often kills by dehydration. Make sure they receive enough liquids, two ml of Vitec stock drench in 100 ml (.4 cup) of water would be safe, but no extra milk. In cases of adults with heavy scouring, minerals are lost from the system and need to be replaced. Drenching with 10 ml (2 teaspoons) of the Vitec liquid, a teaspoon of dolomite and cider vinegar made up to one pint with water, will help replace them. This condition should not arise in animals properly supplemented with ad lib seaweed, and/or the stock lick. In very severe cases an electrolyte replacer may also be used.

In any obstinate case of scouring, presuming it is not due to worms, an injection of vitamin C should be given daily until it stops — four grams for an adult, half the dose for a kid. All this is assuming that the goats are on a tested and remineralized paddock.

Edema

This is caused by the body tissues holding excess fluid and generally occurs in the legs and along the sides of the abdomen. It is generally found in grossly overfed animals, especially if the diet is too high in protein and possibly salt. Years ago goat keepers tended to feed far too much salt to their charges. If they are getting seaweed meal ad lib, salt is quite unnecessary. More exercise, less feed and a balanced ration with all the minerals in it should correct the situation quite quickly.

Edema of the Udder

This is rare and has been mentioned in the section on parsley — usually only elderly does are affected. The udder feels full but cannot be milked. It is engorged with fluid and does not necessarily follow just after kidding. It is *very* uncomfortable. One of the reasons for this happening and it was a contributory factor in this case, was the old practice of not milking a goat before she kidded. This is very cruel and I never could understand why it was promulgated. Does should never be forcibly dried up and, if the milk does come down, milk them as usual. The colostrum does not come down until the kids are born.

Some vitamin C orally will do no harm, but the sovereign remedy if obtainable, is parsley — give the doe all she will eat. I found this remedy in Juliette de Bairacli Levy's *Herbal Handbook for Farm and Stable* and luckily I had parsley growing in the garden (see section on parsley, Chapter 10).

Encephalitis

This is (I hope) very rare in goats. I have only had one case in 35 years and she was *not* CAE positive. No one had any suggestions as to how she could have contracted it and none of the other goats were affected.

When all the goats came in for their tea she was missing; I could see her about half a mile up the farm standing quite rigid. Luckily my son was home, so I got him to drive the Kombi, while I armed myself with a bottle of injectable vitamin C and a syringe. We found her with her legs stiff together under her, her back arched and tail and head up rigid. We got her into the van, and I had given her about 50 cc (in the muscle) of vitamin C by the time we got her back to the shed. She soon collapsed and could neither move, bleat, eat nor drink and I suspected meningitis.

As it was rather late, we bedded her down on her side and made her as comfortable and warm as possible. The vet came out first thing the next morning. The doe was still the same so we gave her 15 grams of vitamin C in the vein of each front leg. Carol, the vet, told her partner that I really had gone mad expecting to save the animal as she had diagnosed encephalitis.

Within two hours of the 30 gram injection of vitamin C the doe gave a bleat — rather a pained one — and managed to drink a little fluid. I offered her hot and cold water, molasses, and then in desperation straight cider vinegar which she drank avidly. The next morning Carol came out and repeated the first day's dose of vitamin C and was very much surprised to find her alive at all. By the evening she was picking at green feed and hay, drinking as she wanted and could now lie happily in the normal position instead of spread out flat on her side.

The last thing that night, when I went out to her, she staggered to her feet and on the next day, although a little groggy, she seemed fine. I continued the vitamin C orally, a tablespoon night and morning, with dolomite and seaweed meal all in one dose for two or three mornings. By the fourth day she was objecting strongly to being away from the other animals so I let her resume her place in the herd. There were no recurrences. She was quite a highly bred British Alpine and lived on as though nothing had happened.

Enterotoxemia (Pulpy Kidney, Entero)

This is caused by the organism *Clostridium perfringens D* and/or occasionally *Clostridium welchii*. Both are normal inhabitants of the gut and only when the goat is under nutritional or other stress — usually worms — do these organisms start to proliferate and in so doing give off a deadly toxin.

David Mackenzie in the first (not updated) edition of *Goat Husbandry* claimed that immunization against entero was not necessary in properly looked after goats. He knew nothing of the often terrible environmental conditions in which goats farmed in Australia are expected to live. But, even so, in goats whose mineral requirements are fully met, changes in habitat or feed — often the cause of an entero attack — make no difference to their health. In fact, many vets have told me in recent years they consider entero to be a much overrated disease in goats because, on the average, goat farmers see to the mineral requirements of their stock. Unfortunately vaccination, two-in-one for entero and tetanus, often only confers a false sense of security to the goat owner. A vaccinated goat can develop entero (or tetanus) just as easily as an unvaccinated one if the conditions are right for either disease. A goat that is dying from some other cause usually is stricken down by entero in the final stages. Vaccination makes no difference to this process.

Kids and young stock are most prone to entero because usually older goats have developed an immunity. Therein lies a large snag. If an older goat that has developed its own immunity — either naturally from contact or from an earlier vaccination — is given an entero vaccination, it may die from anaphylaxis and this will happen very fast. Do not vaccinate all new arrivals as a matter of course, they are better left alone if up in years and their previous history is not known.

Signs of entero are misery and scouring which, if not attended to, rapidly reach the stage where the animal loses the use of its back legs. In advanced cases the scour will contain sloughed off pieces of intestine. Unlike sheep, where entero kills fairly fast, goats always give the farmer plenty of warning and time to take remedial action.

An excellent initial treatment is a quarter of a pint of warm cooking oil; this always seems to be beneficial in any case of bad scouring. Give 10 grams of vitamin C, with one gram of vitamin B12 and two ml VAM in the same syringe, by injection. Then two teaspoons (10 grams) of ascorbic acid powder orally, followed by a heaping teaspoon of each of the following: dolomite, slippery elm powder and crushed garlic tablets. Repeat all except the oil, B12 and VAM at two hour intervals.

I had a vaccinated Angora buck boarding with me many years ago. When I found him in the last stages of entero, after I'd been out for the day, with bits of his gut in the scour and he was unable to walk, I was wondering how I was going to explain his death to the owner. He was very valuable and destined for the sales four days later. I took the measures listed above and by the time I went to bed he was back on his feet taking an interest in life once more. I continued the vitamin C injections for the next three days, but he was back to normal feeding by next day. He was washed and duly presented at the sales on time.

There is an antitoxin available for enterotoxemia; I have only used it three times when goats under my care were dying from capeweed poisoning. In all three cases, the does which had kidded normally before, produced abnormal kids next time. Coincidence or not, it was enough to make me decide not to use it again. None of the kids from other does who were not given the antitoxin showed any abnormalities.

Flag

This is not to be confused with hard udder, a condition that often arises in goats with CAE virus. Some families of goats, especially high milkers, are more predisposed to flag than others. When a doe is freshly kidded, especially if on a diet high in goitrogenic feed (alfalfa, etc.), she could come in with flag. It feels like a hardened ridge running from front to back on the udder where the two quarters meet and spreads out from there.

Cut *all* legumes and green feed out of the diet, including the paddock grazing if high in clover, and feed good grass hay, chaff and bran. Ten ml of Vitec stock drench and four grams of vitamin C powder in the feed would also help. It usually clears up in less than 24 hours. A friend who brought her freshly kidded doe into the Royal Show the afternoon before the milk out for the Q Star test found that the doe had flag so badly that she could hardly be milked. We implemented the above measures immediately and I assured her she need not withdraw the doe from the show, only from the milk test. I don't think she believed me. The doe was shown and was placed in the best udder class the day after the milk test.

For does who have had flag previously, remove the goitrogenic feed at least one or two weeks before kidding. Make sure she receives all her minerals plus ad lib seaweed meal and she will stay free of the ailment. The alfalfa chaff etc., can be reintroduced once she is milking safely.

Also see section on hard udder.

Fly Strike

It is rare for milk goats to be fly struck. Very occasionally scouring animals are struck by small bush flies and the maggots, though extremely small, can still do damage. Any goat showing undue irritation in the anal area should be closely examined.

Fleece goats on the other hand, can be struck in the same manner as sheep, particularly following damp humid weather or a wound, although this is not very usual. Now that the Texan and South African Angora's blood is through quite a high percentage of the national herd it may be another story. Within 48 hours of the first Texan coming out of quarantine in Australia, I was rung about a badly fly blown buck. These animals have a yolk in the fleece, not unlike Merinos and, if they are not well managed, apparently become blown very easily. However, Merinos on the stock lick described in Chapter 6 were struck but, unlike those not having the lick ingredients, the maggots bailed out straight away and no harm was done. Apparently properly supplemented animals do not taste very good.

Any goat showing irritation or misery should be closely examined. In bad cases the fleece will come right away from the animal as the maggots will have eaten it out from underneath.

Trim round the affected area, scrape away as many maggots as possible and apply either a proprietary fly strike lotion — all rather lethal to the handler as well as the maggots these days — or try Flint's Medicated Oils, Septicide ointment or a copper sulfate, vinegar and water wash, rubbing it in. It works well and is useful in this application. Unfortunately, due to the greasy nature of the above remedies and the yellow color of Septicide, these may not all be viable options for fleece goats. Care of a fly struck goat should include watching out for secondary infection on the struck site and keeping it dressed with Flint's Medicated Oils or Septicide until healing takes place would avoid this. If in any doubt give vitamin C, a teaspoon in the mouth daily, until the beast is healed.

As always, prevention is better than cure and, as recounted above, sheep farmers have found the lick minerals work well as a preventative. This would, of course, apply to goats too. Australian blowflies, unlike their counterparts in other countries, lay live maggots not eggs, so fly strike is a far quicker process here.

Foot and Mouth Disease

This is a notifiable disease in Australia (and many other countries). Call the vet immediately if in doubt; most foot and mouth scares here have turned out to be just that. There have only been two outbreaks here as far as I know — one in Melton in the 1880s, and one in Gippsland this century — and both cleared up very quickly. The disease appears to like colder and wetter conditions than those found in Australia.

It only affects cloven-hooved animals and signs are sudden lameness and dribbling, with small vesicles around the feet, between the toes and in and around the mouth area. It is acutely painful and animals lose condition very fast. Foot and mouth disease is endemic in many countries, including some of our nearer neighbors. In Europe birds are usually blamed for its spread.

Anyone who has lived through a foot and mouth outbreak in a country like the United Kingdom where total eradication is the policy, and seen whole herds of cattle slaughtered to prevent its spread, will realize that the disease is best avoided at all costs.

The organism can live 120 days on clothing, possibly longer, so if travelling take great care to have clothing and footwear disinfected on re-entry. Foot and mouth disease is reputedly endemic in the islands to the north of Australia.

Goats seldom contract the disease, possibly because their mineral intake gives them a degree of immunity. In Europe total eradication is not carried out (as it will be here) so that valuable genetic material can be saved. There are documented cases from Europe of cattle who had free access to seaweed meal failing to catch foot and mouth even after close contact with the disease. It is a highly contagious disease. Dettman and Kalokerinos report that it has been cured with megadoses of vitamin C. Thirty years ago in Holland cows on ad lib seaweed meal did not succumb during an epidemic.

Foot Rot

This ailment has nothing to do with wet weather, although the weather is usually blamed for its onset. The organism lives in the soil, especially in areas low in copper. Goats whose copper and sulfur requirements are correctly supplied will not contract it, even if up to their knees in mud for the entire winter.

Foot rot starts with a breakdown in the keratin, the main components of which are copper and sulfur. Keratin keeps the skin nice and waterproof and in good condition. When it breaks down due to deficiency, the first place affected is between the toes, where a thin red line may be seen at the outset. This soon spreads and the feet become sore and smelly. Signs are lameness and an offensive purulent discharge from between the toes (often building up to proud flesh) and under the foot wall. If left unattended for any period of time, the foot may be almost destroyed. Trim away the rotten material and preferably burn it, scrub the affected area with a copper wash made of two tablespoons of copper sulfate and one of vinegar in a liter of water. Put dry copper sulfate on any actual lesions and cover up for 24 hours, by which time they will have cleared off.

Give any affected goat half a teaspoon of copper sulfate mixed with one teaspoon of dolomite and one of vitamin C powder for two days and then make sure that the copper ration in the feed is correct. If it is not the foot rot will reoccur. A lack of copper (see section on that mineral) causes a great many other ailments. It is very important that the requirements for that mineral are met. This is best done by ad lib seaweed meal, and actual copper sulfate where required.

Foot Scald

This is the precursor of foot rot and is due to the same cause — incorrect copper and sulfur levels in the diet. The claims that it is entirely different from foot rot mostly arise because there was a very Draconian response to foot rot which sometimes involved destruction of the animals concerned. Apparently no one nowadays realizes that it is due to a mineral shortfall. Yet every veterinary manual and instruction book before 1960 had no doubt that it was caused by a shortage of copper.

Founder

Founder does occur in goats as well as horses, though not in quite the same form. The chronic variety, where there is build up of tissue on the body as seen in horses, does not affect goats. But sudden attacks of founder have been known to occur after unscheduled raids on the feed bin or other over indulgence. The treatment is the same as for horses, a tablespoon or possibly two of Epsom salts by drench works quite quickly.

The signs of founder are usually sudden lameness, hot and occasionally swollen feet. These clear up once the magnesium levels are raised. Of course in goats that are properly supplemented recovery would be very quick once the Epsom salt (magnesium sulfate) has been administered. Dolomite and the other minerals should be added to the ordinary feed on a daily basis as suggested in the chapter on feeding.

Occasionally goats on a diet that is overly rich in protein will get what amounts to laminitis (chronic founder) where the feet become calcified and chalky making trimming them quite difficult. In acute cases, the animals tend to walk on their knees and avoid using their feet. Remove the cause immediately and bring the diet back to its correct level (see Chapter 6 on feeding). Continue with the regular amount of minerals and gradually the feet will grow out and normalize. It takes a while to happen and cannot be undone in days.

Hereditary Defects

The more obvious defects like cleft palate, over- or undershot jaws, teat, udder and scrotal deformities are all well documented recessives. This means that they will be "carried" by both parents and possibly only appear rarely. A mating that produces such a mistake should be noted and not repeated. These recessives have been with us a long time and sometimes it is equally long before they become apparent; often one has to go back 10 to 15 years in a pedigree to find the original culprit.

Possibly, if everyone culled any goat bearing such faults or refused to keep kids from them, recessives might be wiped out, but it is problematical — the small gene pools that we have in this country make eradication impracticable. Fleece breeders often find small discolorations occurring "out of the blue," so avoid the mating that produced it. In those cases it might be nearly impossible to find where the fault originated.

In the early 1960s deformed and double-orifice teats were running at almost 80 percent in the British Alpine breed (the usual incidence in any breed as about two to three percent). This huge jump was caused by a buck that was kept from a doe with the fault — which no one had noticed. Rigorous culling, in a breed that could ill afford it, eventually brought back the incidence to an acceptable norm.

Malformed Scrotum

This is a very serious fault, especially in dairy goats, and cannot be tolerated in any future stud buck. Testicles that are uneven will often denote a fairly high percentage of uneven udders in the buck's female offspring.

Undescended or missing testicles are also a bad fault and their bearer should be culled. Note that in the case of undescended testicles castration is not possible except surgically.

Malformed Udder

From 1975 to 1988 there has been an absolute epidemic of one-sided udders in the dairy industry and possibly some good goats were unjustly penalized. Once the CAE virus was brought under control, one-sided udders ceased to be a problem as suddenly as they had started. Reports

Good rear end and udder

Cow hocks and divided udder

Well attached scrotum

Poorly attached scrotum — pendulous and uneven

Udders and scrota.

from the United Kingdom first alerted Australian goat farmers to the fact that it appeared to be caused by CAE. However, should this occur in a healthy CAE-free doe, it should be treated very seriously and she should not be used for breeding.

A one-sided or very uneven udder in a milking doe has nothing whatever to do with a goatling (unkidded) that develops one side of her udder, or a very uneven one. This should be watched to see it does not become too tight — in which case it should be milked out. When these goatlings become milkers their udders are generally even.

Anyone who has hand milked large numbers of goats will know that most goats show a slight bias to one side or another; however, that cannot be construed as an uneven udder unless it is palpably so both before and after milking.

Spermiostasis and Cystic Ovaries

Spermiostasis can be hereditary, in which case it is found in heavily polled families where the males also have a predisposition to spermiostasis. Some authorities claim that any male that is pure for poll will throw spermiostasis in his sons and cystic ovaries in his daughters. I had the daughter of a buck which was pure for poll who did have mild spermiostasis (which allowed him to successfully cover three does). That daughter lived to a ripe old age and never showed the slightest sign of cystic ovaries.

Both conditions cause sterility, either straight away or later on in life, and are therefore uneconomic. A buck with spermiostasis is always polled and may possibly, as in the case above, cover a few does before the pressure from the blocked ducts in the scrotum sets up putrefaction and the animal becomes completely sterile. In Europe between the two World Wars there was a great prejudice against horned milking goats and they could not be registered. All goats were bred for the polled factor and horned kids generally destroyed. Inevitably the goat population, particularly Saanens in Germany, became almost pure for poll and, just as inevitably, the rates of spermiostasis and allied conditions reached 50 percent of the goat population. Then, in an attempt to avert a disaster that had already happened, it was decreed that only horned goats should be kept. Nowadays people are more rational and realize that while polled goats do tend to be the best milkers, they must be bred with horned animals at intervals to avoid infertility problems.

Except in very mild cases spermiostasis may be detected by palpating the scrotum as the hardening along the top of the sperm ducts may be felt. A veterinary surgeon will confirm your diagnosis. In cases of infertility, where the hardening is not apparent but the buck is polled and from polled parents, a post-mortem usually reveals the early stages of the malady.

See diagram of scrotum showing seat of spermiostasis.

Cystic Ovaries

In goats, unlike cows, these cannot be felt by palpation — there is no room. They can only be detected by the behavior of the doe (or on post-mortem). Signs are masculine behavior with the doe apparently partially in season; these animals can become quite aggressive and have been known to cause actual injury to other does. They do not stand to the buck very often and, if the rear end is examined, it will be found to be pillar box red instead of the normal mildly swollen pink. Post-mortem examination will confirm this condition. Cystic ovaries can be present from birth, but they can also develop after several kiddings — in my experience, this only happens to animals on inadequate dietary minerals.

Wry Face and Wry Tail

The former has already been discussed in the chapter on kids. It is a disqualification fault and is apparent from birth. If an older animal develops a twisted face, it can be due to disease (such as pneumonia affecting the facial bones) or to malnutrition.

Wry tail is often undetectable when a kid is born; it can be a result of the tail being turned back *in utero*, which straightens in a few days. Nubians have it (legally) quite often. It is believed to be allied to a crooked spine by some authorities. Obviously, unlike a crooked face which may affect an animal eating, it is not so serious, but it is a disqualification fault in the show ring. It possibly would be unwise to keep a buck showing the deformity, but it will not affect the milking ability of a doe.

Intersexes

These are kids that are neither one sex or the other. There are variations and generally, genetically, they are males even if this is not apparent. Occasionally they have both scrotum and vulva, often misplaced. All kids should be closely examined at birth, if in any doubt consult an experienced goat keeper or vet. They do not make good pets. Their disability seems to affect them mentally and make them unpredictable and the misplacement of their organs often makes them dirty. (See illustration of normal and abnormal clitoris, Chapter 8, page 133.)

Ill and Unthrifty Kids

These have already been mentioned in the section on kids. Do not go to great lengths to save kids that are sickly at birth unless there is a really good reason for it, such as the doe being hurt or chased. Nature usually knows what she is about and does not intend them to live.

Impaction

This is another name for constipation. The animal strains but does not pass any feces. When and if it eventually does pass feces, they will be hard and dry. A pint of good vegetable oil for a large goat, less for a kid, will help. This should be followed by a teaspoon of ascorbate (vitamin C powder) as a drench, which also has a mild laxative effect and would restore the health of the gut.

Never use liquid paraffin as it demineralizes the body because this would be especially dangerous for kids.

Immunizations

These cause an almost total depletion of vitamin C in the system, so it would be wise with valuable animals to give a little extra of that vitamin before these procedures. This explains why some animals (horses in particular) become ill after routine vaccinations.

Injury

Call a vet if possible, especially if there is badly torn flesh. All wounds must be thoroughly disinfected straight away. Any good germicide will do and it is important to do a thorough job as the initial disinfecting should be the last. Years ago vets taught me that disinfectant of any sort inhibits healing and should not be used more than once.

Both tetanus and blackleg are the result of uncleaned (often unnoticed) wounds. Fresh, clean cuts can be disinfected quite easily and then stitched with an upholstery or surgical needle and linen thread, both properly disinfected. The goats I have done showed absolutely no discomfort, and did not appear to feel what I was doing. If they do, try to get a vet to give a local anesthetic.

If a vet is unobtainable, after disinfection tidy up badly torn wounds as best you can. Put on a packing of comfrey ointment if possible, otherwise aloe vera — Septicide and Savlon are good proprietary lines. If none of these are available, use a dressing thoroughly soaked in Flint's Medicated Oils and bind up the wound. If the wound is deep and has not bled much, tetanus or blackleg will be a real possibility. The former takes around 10 or more days to come out, the latter three or four. If a vet is available the goat will have been given antitoxin for both, otherwise keep the animal on an elevated oral dose of vitamin C for the 10 day period, a dessertspoon a day (about 10 grams) would prevent either disease and any other condition that might arise.

Once the wound is safely bandaged give four grams of vitamin C by intramuscular injection, as well as some form of supplementary vitamin E — about 2,000 units either by dissolved capsule or injection (intramuscular). Continue with the oral dose as suggested above. If there is any sign of blood poisoning, such as heat around the wound or elevated temperature, resume the injections of vitamin C until they disappear.

Johne's Disease

In the past years of CAE, many goat keepers have become accustomed to unthrifty goats. Often farmers did not realize that it could have been Johne's (caused by *Mycobacterium avium* subsp. *paratuberculosis*) in addition to CAE or on its own. Goats with Johne's have variable appetites; occasionally they scour with bubbles in the droppings and eventually they die from starvation, their food apparently doing them no good at all. The condition causes a thickening of the intestine walls, effectively preventing the animal from absorbing any nutrients from its feed so it virtually does die from starvation.

The main mineral needed if Johne's is to be avoided is copper. As long as goats are being properly fed it is neither a risk or a possibility. In spite of oft repeated warnings of the extreme contagiousness of the disease, in really well kept herds where the diet is minerally balanced, the accidental introduction of a case of Johne's seldom has a recurrence. David Mackenzie says the same and, as in France, where the condition has been fairly linked with inadequate diets too low in phosphates, he was probably correct.

It used to be thought that it was contracted at birth only. However, this has had to be reassessed.

Some years ago, a breeder set up with top-class goats from Johne's-free studs. His wife did not like goats and when he was away on business, which was quite often, they were not looked after at all. In a comparatively short time Johne's Disease was endemic in that herd, every animal that was sold in the eventual (and inevitable) break up was infected with it.

Knuckly Knees and Pasterns in Newborns

See bent leg and contracted tendons.

Lactation Tetany

See tetany.

Laminitis

See founder.

Leukemia

I have only seen three cases of leukemia in goats. Strangely, all were brought to me within a period of two weeks in the late 1960s, each from different districts and studs. We did not know what was wrong, the only signs were large edematous swellings which went from the jaw to chest. I took all three down to the Melbourne Veterinary Clinic, my veterinary support at the time, and, after blood tests, leukemia was diagnosed. They wanted to put all the goats down, but as I did not own them I could not give permission and took them home again.

Three goats, all with one factor in common. Number one had come from a small farm in Warrandyte next to a rose farm from which the sprays drifted over continuously. Number two came from a farm that had just bought in some alfalfa hay for the milkers and it had been sprayed (incorrectly) the day before it was harvested. Number three was a beautifully kept backyard goat belonging to a very dear friend (I wanted to try and save at least hers) who fed her largely on outside vegetable leaves obtained from the local greengrocer. The major part of the sprays would have been on the leaves. None of this was known to me when the goats paid their first visit to the vet clinic at Werribee, Australia.

I kept them at home, fed them normally on a good plain diet and they grazed my organically farmed goat paddocks. Three weeks later all were on the road to recovery and, after four weeks, I took them back to the vet clinic in Werribee appearing quite normal. The blood was tested again and they were clear. My suggestion of sprays being the only common factor was not met with much enthusiasm. No other reason was forthcoming and leukaemia was deemed incurable. These days there would have been no skepticism, sadly we are wiser now — sprayed produce *must* be avoided.

Lice

A lousy goat of any breed shows rather bald ears and spends much time scratching itself on fences or anything else in reach. In dairy goats halting an infestation is not much of a problem, but with fiber goats, especially in full fleece, it is another story.

As always, prevention is much easier than cure and animals receiving the correct amounts of sulfur in their diet will not get lice. See the section on sulfur for the reasons behind the deficiency. For short-haired goats (who usually catch the lice from a boarder), give one-and-one-half teaspoons of sulfur per head per day to all the goats in the shed and keep that level up until the lice have gone. Go they will — where to, I have never discovered.

In very severe cases the animal will be eased by rubbing the sulfur along the back line and underneath around the legs. The animal should clear up in about 24 hours. Pesticides cannot be used on milking goats as it comes through in the milk within 24 hours. The pour-on variety in particular has caused abortion in goats — so beware.

For an infestation in fiber goats, feed the elevated sulfur; it will stop the lice living close to the skin and hopefully they will die. Once the goats have been shorn, care should be taken to keep the sulfur levels up so reinfestation cannot occur.

Sulfur can be included in a lick offering. For hand-fed milkers it is best included as a maintenance ration in the feed — a rate of a teaspoon per head per day is usually enough — it must not exceed two percent of the feed, which allows fairly high quantities to be fed. On farms where lice are endemic, serious thought should be given to remineralization, spreading gypsum (calcium sulfate) with the dolomite or whatever.

Liver Fluke (*Fasciola hepatica*)

Signs of fluke infestation are ill thrift, occasionally with a lump under the jaw, anemia and variable appetite. If a postmortem is performed, the fluke, which are small flat- (fluke) shaped objects, will be found in the liver. The fluke has a six week life cycle; therefore, if using drugs to counteract them, they must be given at a six-week interval.

Any farmer who runs cattle on irrigation — one of the best ways of spreading the host snail — will tell you that copper sulfate is the answer. Having farmed goats on an irrigation system where fluke were endemic, I also found this to be the case. The maintenance dose of copper sulfate that my goats habitually received (one teaspoon per head per week) was enough to prevent all infestations.

A small conical snail is the intermediate host and sometimes, if the infestation comes from a dam or very soggy pastures, they can be seen on the edge of the water or on the ground. On a farm in Gippsland, Victoria, Australia, in a very wet winter the goats acquired fluke; I was not feeding enough copper to cover the inherent deficiency in the food and the completely copper-deficient pasture.

If there is an infestation, the drugs are very expensive and vets often do not carry them for that reason. When my vet told me this, I said I would raise the copper in the diet as it had prevented fluke on my previous (irrigated) farm. He was quite horrified and said that it would damage the liver. I reasoned that the fluke were doing just that fairly effectively anyway and went ahead. For seven days each goat was given the equivalent of a teaspoon of copper sulfate daily run through the feed. I watched them go from a state of listless anemia to bright, healthy looking goats once more. The next goat to receive a post-mortem a week or two later had no fluke at all and she had looked the worst. (In the CAE era, performing post-mortems on goats was the price we paid.)

If you suspect a dam or waterway is the cause of infection, examine it carefully and if you find the snails, take some to your veterinarian to check if they are the fluke variety. It is a matter of which way the spirals run on the shell. If they are flukes, throw half a kilogram (1.1 lbs) of copper sulfate into a normal small farm dam, a little more might be needed on a big one. It is difficult to treat a waterway, but farmers on irrigation tell me they put a small, thick canvas bag of copper sulfate at the edge of the incoming stream of water, which has the same effect.

A basic maintenance dose of a teaspoon of copper sulfate per head per week for goats is effective in stopping fluke infestations. Smaller amounts could be tried and the goats regularly tested, but I have found the teaspoon a week to be the best. Fleece and meat goats do not need as much as a teaspoon per head, half this amount should be enough. The copper in the lick offering free-choice as described earlier should be fine.

Lungworm

See worms.

Lung Damage

This can easily be diagnosed by testing the goat's breathing rate, it should be 20 to 24 per minute. An adult doe that, following pneumonia, had a breathing rate of 120 (per minute) from lung damage was given 1,000 units of vitamin E daily. After 10 days of this therapy, the breathing rate was reduced to 35 to 40, the lowest it would reach, obviously a considerable improvement. This might possibly have been bettered had we known about using vitamin C as well, which also has a healing action — a dessertspoon (10 grams) orally in the feed each day would have been sufficient. This treatment can be tried after severe lungworm damage.

Goats, as they are not an animal that lives by speed like a horse, have a small lung area for their size, so any lung damage is crucial and if it occurs they are left with very little operative space.

Mastitis

There are several types of mastitis, all caused by different organisms, *but* in a doe that is being properly fed (i.e., receiving her dolomite and minerals in their correct levels regularly) mastitis should not occur. Too high protein in the diet can be a causative factor, this depresses the copper and when that happens the immune system does not function as it should. If the protein in the food is excessive, lower it.

If a doe persistently becomes infected with mastitis, have her tested for CAE. It is, sadly, the usual reason for the inability of the immune system to do its task. She is incurable, except in the short term, if that.

All mastitis appears to be due to an imbalance in the health of the udder, particularly the pH, caused by incorrect calcium/magnesium ratios in the diet. Cows and goats that are regularly supplemented with dolomite in their ration stay free from the disease. Many cow dairymen as well as goat farmers have found this, as did I, from the first time I was told about dolomite in the late sixties. For reasons as yet unknown, low levels of these minerals place the udder at risk for invasive bacteria. Diseases only occur when the food is unbalanced and missing the proper nutrients.

There is no doubt that badly adjusted milking machines can be a causative factor, but the cowmen to whom I have suggested using dolomite used milking machines. They said that mastitis (and acetonemia) became a thing of the past from the time they started to feed it. As stated, feeding a diet too high in protein can also be a causative factor — check the section on feeding.

I ran up against this with English commercial goat keepers. I had forgotten how good the soils were over there and the fodder on offer had double the protein we would get here. Extra copper added to the ration of any goat getting mastitis made the dolomite and vitamin C work, but on their own they did not.

The advantage of using vitamins and minerals to cure mastitis is that the goat does not develop drug resistance — nor is the farmer left with a goat with a wrecked udder, as often happens when drugs are used. When an antibiotic is used on mastitis, a new one has to be found next time. I was told to destroy all my oldest goats with subclinical mastitis as they were incurable, which was when I had to find an alternative because my goats were too valuable.

I learned about adjusting the calcium/magnesium balance originally from Mrs. Maura Mackay who used to run a goat dairy before they changed over to breed the famous Glenroy Angoras. I used dolomite alone to cure many stubborn cases of mastitis in the early days as some of the goats I originally bought had advanced staph mastitis. I was informed by the sellers that lumps in the udder were hereditary. When I first learned to use vitamin C as well as the dolomite, I found that the cure worked considerably faster.

Long-standing cases where the udder is a total wreck are usually beyond anyone's powers, particularly with organisms like Klebsiella, which have usually moved in by that stage.

One doe I bought, who had not been milkable for three years because of staph mastitis, became a useful member of the herd after four weeks of treatment. She was given a dessertspoon (2.4 teaspoons) of dolomite and vitamin C powder daily, as well as the routine dolomite and other minerals in her feed. I also gave her a course of five grams of vitamin C by injection for the first three days.

The alert farmer may nip mastitis in the bud by being observant at milking time — a doe who kicks and makes an unusual fuss at milking should be carefully checked.

Kinds of Mastitis

Black mastitis, Streptococcus, Staphylococcus, *Mycoplasma agalactiae*, Klebsiella, and possibly CAE as a predisposing factor in some cases, are some of the organisms that can invade an *unhealthy* udder and cause mastitis.

Black Mastitis

This is the term for a very sudden and severe attack of mastitis. In a matter of an hour or two the whole udder will become contused and, unless immediate action is taken, it will be wrecked and the goat will die. In such a case the doe may be alright at morning milking and be *in extremis* five or less hours later, often following a wound. The udder is hot, hard and inflamed and the goat is obviously very ill with a high temperature. In severe cases that are not treated immediately, the udder will turn a greenish color and slough off (if the doe lives long enough).

Quick action is essential. Five grams of intravenous vitamin C first, if possible; if not, give intramuscularly with a heaped teaspoon of dolomite and the same of vitamin C orally. My cattle dairy farmers are using six percent pharmaceutical-grade hydrogen peroxide for this complaint and one of them told me he cured a cow with black mastitis the same day. A goat would need three ml of hydrogen peroxide straight into the teat orifice, diluted 50 percent with rainwater. *If* this works as well as it did with the cattle it will be a breakthrough because one has to be *very* quick to cure this complaint.

Repeat the entire regimen described above in one hour, then every three hours for the first day and again the second day if no improvement is seen. Otherwise continue to give the oral dose night and morning, with five cc of vitamin C by injection daily until the udder starts to look and feel normal. Cease the injections, but continue with the oral dose of vitamin C and dolomite daily until all lumps are gone. This usually takes about ten days. During this time, milk the udder out as gently and as much as possible. The doe will need careful and patient handling.

Clinical Mastitis

This is usually due to a Streptococcus infection. The milk becomes viscous, stringy and offensive. Treatment as above will bring about a recovery.

Subclinical Mastitis

This is an insidious complaint usually caused by a staph infection and is difficult to diagnose. The first sign will be that the milk, instead of keeping the usual seven to 10 days, will "go off" by the third or fourth day which clean, properly cooled milk from healthy goats does not do. The test for subclinical mastitis cannot be done with a rapid mastitis check, it has to be plated for at least 36 hours for diagnosis. If the condition is allowed to continue unchecked, round, hard lumps will start forming in the udder and eventually the doe will become ill and the udder useless.

A heaped teaspoon each of dolomite and vitamin C powder night and morning for three days, or as long as it takes for the milk to be tested clear, will effect a cure. The lumps, if long standing, may take a week or two of this treatment to disperse.

Mastitis Caused by Avocado Foliage

It is only in the last few years that farmers have realized that this can happen to both cows and goats. Initially, the avocado reduces the amount of milk quite materially and, if the beasts are not prevented from grazing it, mastitis follows. Do not allow goats (or other lactating animals) to eat avocado trees.

Procedures

In all cases of mastitis, milk out the affected animals as usual, but milk them after the healthy animals and take strict hygienic precautions. The mandatory dose of "dry cow" when a goat goes dry (*if* she does) is quite unnecessary if the does (fresh milkers as well) are fed their normal amount of minerals regularly as explained in Chapter 6. This, of course, includes cider vinegar fed on a regular basis; it is another great help in complete udder health, 10 to 20 ml a day per goat.

Metritis

This is an inflammation of the uterus and quite often the only sign is slightly lowered health. It can occur at any time, not always just after kidding. Sometimes a slight offensive discharge is seen, but more often it is lowered health that alerts the farmer.

Metritis should not occur in healthy animals. In recent years it seems to have been confined to CAE-positive does. A vet will confirm the presence of the disease. If it occurs out of the breeding season, the vet will possibly say the same thing that I was told when one of my goats first contracted it: "Feed her up well and she will get over it." I did just that and by the breeding season the swab was clear (this has to be done when the doe is in season).

However, a base lack of the necessary vitamins A and D is the chief cause for all diseases of this kind. Extra vitamin A in the form of a teaspoon of cod liver oil orally for three days, with a dessertspoon (2.4 teaspoons) of vitamin C orally for the same time, will usually make certain the next swab is clear. However if the doe is very ill, five grams of vitamin C by injection should be given daily for three days as well, followed by a dessertspoon (2.4 teaspoons) of the powder orally for at least a week. Continue with the vitamin A and D for the same time. (If does contract metritis out of the breeding season, it may well be seven months before they can be swabbed). These uterine disturbances often mean that the doe's diet has been low in vitamin A — due to drought or poor quality feed. A long dry period puts animals who do not get their cod liver oil occasionally at risk.

Any doe that has had metritis whether on the farm or coming for service *must* have a clear swab (done when she first comes into season) before going to the buck again (see section on infective arthritis).

Metritis can make milk unsafe to drink and it must be heat treated before being fed to a kid (goat) or human.

Milk Fever

This is not really a fever, rather the reverse. The doe will be low in spirits to the point of lethargy following kidding. She will have poor muscle control, difficulty in standing at all, her pupils will be enlarged because the eye muscles have relaxed — all very similar to snake bite.

The sudden drain of calcium and magnesium from the doe's system following kidding will mean there is not enough to sustain her. Milk fever does not seem to occur in does who have been receiving the correct minerals.

Quick action is necessary or she will die, give calcium borogluconate (with magnesium) or any proprietary milk fever preparation as per directions on the bottle. This is easily obtained at any farm store. The injections can be given in four doses to speed the process. Give injections to each side of the shoulder and each side of the rump. *Hungerford's Diseases of Livestock* states that this injection must include both calcium *and* magnesium.

Injection sites for milk fever remedies.

Some does are more prone to milk fever than others, usually (but not invariably) the high producers. It would be wise to give them a little extra dolomite coming up to kidding and a vitamin A, D and E injection.

Mycoplasmosis

See pleuropneumonia.

Nasal Bots

This bot is very like the larval stage of the bot fly which affects horses. It is pale chestnut in appearance, with an oval body and a distinct head — roughly one-half-inch long. The fly lays its eggs just inside the goat's nostrils, the tiny larvae crawl up into the animals head and live in the sinuses, nasal passages and occasionally reach as far as the brain (see circling disease).

The first sign will be much sneezing among the goats. Eventually, when the larvae are full grown, they will be blown out and the pupae hatch and repeat the cycle. Vicks, K7 or similar substance smeared around the edge of the goats' nostrils can sometimes make the goat sneeze the larvae out before they are ready, or it may discourage the fly from laying the eggs initially. There is not much that can be done, except to hope that the bots do not reach serious proportions as often happens with sheep. There is a suggestion that systemic worm and lice/louse preparations may kill the bots. This would not expel them from the body, which at least does happen when they follow their normal cycle. Fortunately nasal bots are quite rare; there may be an outbreak one year but it is not usually an annual event and lasting damage is unusual in goats.

Maggot

Bot Fly

Nasal bot fly and larva.

Ophthalmia, Conjunctivitis, Pinkeye

When this condition (also called Sandy Blight) is present, the eyes water excessively — (which can also be due to a vitamin A shortfall) and become opaque. Eventually, if no action is taken, the whole eye becomes bloodshot and may rupture causing permanent blindness.

This so-called disease appears to be wholly due to a deficiency of vitamin A. Supplementation with that vitamin, accompanied by topical application, brings about a cure even when the eyeball looks beyond redemption. If the pink eye is not treated for a week or so, the sight often will not return for the full three weeks, but if the rupturing stage has been reached the eye may be damaged.

Of course the blinded animals must receive particular care. Give vitamin A, D and E injections as per instructions, but better still fill a 10 ml syringe with cod liver oil and squirt two ml into the affected eye and the rest down the throat. This topical application is extremely effective. Another remedy which is of great help in removing the soreness is to obtain a teaspoon of Ferrum Phosphate powder (from crushed biochemic tissue tablets) and gently blow it into the open eye.

A farmer whose farm was halfway to organic conversion had some Texas Longhorn bullocks contract pinkeye when in an unrescued paddock. As they were rather large to treat manually, he transferred them to an organic paddock of green alfalfa and the pinkeye cleared up in a few days. Since the disease is caused by a lack of vitamin A, it should not occur in paddocks that have been remineralized.

The above suggestions have frequently worked after unsuccessful treatments with conventional drugs. Pinkeye is very contagious (I remember being incarcerated for three weeks at school with it), but only to vitamin A deficient goats and it is usually spread by flies.

Pinkeye should not be confused with eye damage. This is quite different, the eye may cloud over as the result of a blow or scratch. Careful examination will often reveal the cause of the damage to be a slight dent or scratch on the eye. However treatment similar to that for pinkeye is a good idea for a few days to prevent any further deterioration. Damage usually takes three weeks to heal and the cloud on the eye disappears.

A good wash for eyes can be made from a teaspoon of borax in 2 cups (a half a liter) of warm water. Grass seeds can also cause very sore eyes, here the topical treatment with cod liver oil works especially fast.

Ostertagia, Brown Stomach Worm

See worms.

Parasites (Internal)

See worms, coccidia and liver fluke.

Pinkeye

See ophthalmia.

Pin Worm

See worms.

Pleuropneumonia (Mycoplasmosis, Mycoplasma, Pneumonia)

This disease nearly destroyed the cattle industry in the early days in Australia. Its advent in the goat world was complicated by the onset of CAE and quite often the two were obviously contracted together. CAE could not be diagnosed at that stage, as no one knew what it was. Many a researcher was completely baffled at what looked like a certain case of mycoplasma which yielded no bacteria at all.

The signs are so similar to ordinary pneumonia that they can be indistinguishable. The aftermath is really the only criterion for certain diagnosis without a vet doing tests. In ordinary pneumonia (see next entry), good nursing will bring a goat through and it will not recur. In mycoplasma pneumonia good nursing will again prevail, although the goat may be left with an elevated breathing rate and damaged lungs, but the attacks will keep recurring. The heart is damaged and eventually the goat becomes less thrifty after each attack until it dies.

The condition, like pleuro in cattle, is very contagious. This is a good reason not to house animals too closely in poorly ventilated places; it only takes one sneeze and the organism is carried by the droplets. In kids it often shows up as a chronic runny nose, which again affects others upon contact. Kids and all goats should be kept in well ventilated sheds that are dry and fairly draftfree.

The only drug that has been at all successful for mycoplasma is Erythromycin; the ordinary Terramycin which was much used in the past apparently did not work. Large doses of vitamin C with Erythromycin make it work even better according to modern medicine.

However prevention is the key and, as mentioned, respiratory diseases and a lack of calcium and magnesium have been documented. The inherent deficiencies in those two minerals probably would provide an explanation for the pleuro outbreaks here in the early days. No animals should be herded into sheds where the spread of disease could happen so easily, always avoid situations where this could happen.

Pneumonia

Again this disease should not occur, and once I had learned that goats needed their dolomite, sulfur, copper, etc., it did not figure again. Back to good management.

In this disease the lungs become progressively more waterlogged (hence the name) and, if allowed to proceed unchecked, the goat will ultimately die from lack of air. Sudden fever, nasal discharge, rasping and quickened, noisy breathing — the rate much increased (normal rate is about 25 breaths to the minute) — and a high temperature are signs of the disease. If one listens to the chest just above the elbow, it sounds rather like an express train passing through a tunnel.

Good nursing is the key. Keep the patient in a warm and slightly damp atmosphere by making a tent over the top of the goat's stall — a vaporizer as used for children would help. Give eight grams of vitamin C, one gram of vitamin B12 and one gram of VAM by injection, followed by three grams of vitamin C every hour until the crisis is passed. Give warm water with organic unpasteurized cider vinegar added (about a cup to three quarts/liters). A few leaves of horehound can be added to the hot water if obtainable. I have found that its age-old reputation for helping lung complaints is quite justified. Once the goat is a little easier, halve the injections and supplement with a teaspoon of oral vitamin C powder (five grams) and the same of dolomite, plus a teaspoon of cod liver oil. Garlic in some form (four to five tablets, a clove/corm or whatever is available) daily would also help.

As soon as the goat will eat offer fresh branches and green feed. Bran and alfalfa chaff is usually the first feed that an ill animal will consider. *No* grain should be given. If the goat is slow in regaining its appetite repeat the VAM and vitamin B12 injection (combined in the same syringe). Allow the animal out on the best grazing possible, with ample sunlight (and shade), and rug it if the weather is at all cold.

Usually pneumonia leaves lung damage in the form of scarring and a course of vitamin E as suggested in the section on lung damage would be beneficial.

Poisons

Poisonous plants have been covered in Chapter 6. Other toxic materials are occasionally met on the goat farm.

Arsenic

I have nursed a goat through this form of poisoning. She was vomiting and it is rare for a goat to do that and live. Her breath had a strange smell and she was in a state of total collapse. I used everything I knew, massive intramuscular injections of vitamin C (10 grams), vitamin B12 (three ml), vitamin B1 (three cc), vitamin B15 (three ml), and vitamin E (2,000 units orally). A tablespoon of vitamin C and the same of dolomite were also given orally with slippery elm. Both of these have an absorbent action in the case of toxins. All this was given every hour for the first few hours until the extreme signs abated — it took 12 hours of fighting before I realized that she would live. A hollow victory, because further enquiry elicited the fact that arsenic causes chromosome, bone marrow and possible renal damage and, in spite of all efforts, the doe was never very strong again — but one has to try.

Fireweed (Senecio spp.)

This is generally found in the northern parts of New South Wales and invariably grows on very poor, played-out land. A farmer who had top dressed his fireweed-infested farm three years ago in northern New South Wales called to tell me it was all gone the following year — much easier and more effective than pulling it up. He spread the lime, gypsum and dolomite as advised on the soil analysis.

Goats would not normally eat fireweed unless grazing was very scarce, which it usually is if the land is poor enough for it to proliferate. There is no sign of any malaise for about 18 months. The doe I had kidded normally and a few months later she started to go downhill. I was fighting with everything I knew; the vets had no clue as to how to treat her either. Then I went up to Queensland to talk at a goat seminar and Dr. Ross Mackenzie was one of the speakers; he spoke on poison plants (see the excellent book he and Ralph Dowling wrote, *Poisonous Plants; A Field Guide*) and he mentioned the poisonous action of fireweed and its effects on animals. The description so exactly fit my doe that I had a talk with him and said she had come from northern New South Wales. He told me to post-mortem her when I got home and that I would find the liver had become, small, flat and hard and the edges would have a scalloped appearance. He was right and I kicked myself for having let her suffer.

Nitrate Poisoning

The classic sign of this sort of poisoning is a strange sweet smell on the scour. It is quite unlike the smell of normal manure or of a scouring goat with intestinal disturbances — and only occurs with nitrate poisoning. Motor disturbances, such as convulsing at sudden noises, could easily cause the condition to be mistaken for tetanus in the first instance, as was done by the vet and myself in the first case I had.

As mentioned in Chapter 6, Dr. Selwyn Everist said that vitamin C was the only remedy he could suggest, but it does not always work. An initial injection of five to seven grams followed by a teaspoon of ascorbic acid and dolomite powder, a drench of 10 ml of Vitec liquid seaweed should be given immediately. I found that the poisoning had set up a long-term fatal iodine deficiency. The deaths stopped once I made iodine available in the form of ad lib seaweed meal.

On post-mortem the blood appears a black color due to lack of oxygen, I recommend to have it checked by a vet.

Organophosphate Poisoning

This poisoning is the most dreaded by any vet or doctor because there is so little that can be done. According to Dr. Kalokerinos, vitamins C, A, E, and zinc are the best antidotes for humans. This saved some alpacas willfully poisoned in New Zealand some years ago. The latter could be introduced in the form of seaweed meal in small quantities as it contains high zinc. Give a teaspoon a day and let the goat take more if it wants it. For one week provide daily doses of 60,000 units of vitamin A, 20 grams orally (two tablespoons) of vitamin C orally and 10 cc daily by injection, and 2,000 units of vitamin E either orally in dissolved capsules or by injection. This poisoning would be very much a case of "playing it by ear;" it will depend on good nursing and offering any feed that the animal would take — no grain in any form — just bran, alfalfa chaff and green stuff, depending on the rate of recovery.

Phosphorus

Phosphorus is found in some rat and vermin baits and produces a sweet smell on the breath accompanied by a craving for water. *On no account must the goat be allowed to touch any liquid* — the phosphorus needs water to activate its lethal effects which burn the intestines away. If the goat has drunk, shoot it as quickly as possible or it will die in awful agony. Give the goat egg whites (six at a time) mixed with a little glucose every hour by mouth and by injection provide five cc of vitamin C and one cc of vitamin B12 every three hours. These must be given until the animal shows signs of relief. This procedure may have to go on around the clock for 24 to 48 hours (it took 36 hours with a dog under the instructions of a vet). The burning sensation in the gut makes the animal stretch as though trying to cool its abdomen. Once it is recovered a drink of milk and water, about two pints all together, may be given and gradually the goat can be reintroduced to bland feed and green stuff.

Prussic Acid (Also see Chapter 6)

The antidote to prussic acid is a neutralizing substance such as pharmaceutical chalk or fine dolomite. Either should be mixed with water and drenched in — a tablespoon of powder in 200 ml (.85 cups) of water, both work equally well and fast. This poison is most usually found in young sugar gum shoots and occasionally in wilted peach tree leaves.

Poison Baits

If the constituent is unknown (but *not* sodium fluoroacetae 1080) proceed as for arsenic poisoning. Once the animal is stable give 10 cc of vitamin C by injection and 1,000 units of vitamin E (consult the bottle) by injection daily. A dessertspoon (2.4 teaspoons) of sodium ascorbate and bland feed, such as bran and alfalfa chaff, as well as branches and good grass — all will aid in recovery.

Slug Bait (Metaldehyde)

Goats should not usually have access to slug bait. However I had a case where a goat got into the garden and ate an ice cream container of bait. I had chased her out and did not know she had eaten it until after she was cured. That evening she came into milking looking very ill indeed and, having no clue as to the cause, I gave her 10 grams of vitamin C by injection. She looked much better the next day so I gave her half the initial quantity and then noticed that she had a row of bumps down her spine. These came up as large boils so I continued the 10 grams of vitamin C daily until they cleared up and burst. She recovered fully and then I found the empty bait container and realized what had happened. Since then *no* poisons whatsoever have figured on my farm.

1080

This poison is made up of 23 ppm sodium fluoride (sodium fluoroacetate). If the antidote, which is glycerol monoacetate, is not given within 20 minutes of ingesting the bait — shoot the goat. When carrots are used as a baiting medium and birds pick them up and drop them, goats taking in 1080 is a real possibility.

The animal will live for three or four hours after taking the bait and die in terrible pain. The antidote cannot work after the initial 20 minutes. Vets do not, as a rule, carry the antidote because it is expensive and apparently does not keep indefinitely. The vet who investigated this for me after we tried in vain to save a neighbor's dog, said it was also very difficult to obtain. I am told that 1080 does not cause pain, both the vet and I would seriously query this; the dog died in her surgery.

Fluoride in Reticulated Water (Sodium Fluoride)

This substance has an enzyme-inhibiting action. This is caused by fluoride rendering calcium and magnesium unobtainable in the body. Without magnesium the enzyme system cannot function according to a paper recently printed in the *Townsend Letter for Doctors and Patients* from the United States. Fortunately this does not too often affect goats, although I have read of one herd that was quite unwell until they were taken off fluoride treated reticulated water.

Pregnancy Toxemia

This condition usually arises shortly before kidding. It is due to the kids *in utero* (usually multiples) taking all the nutrients (mainly minerals) available and eventually leaving the doe insufficient to sustain life. Nature always gives priority to the kids — the mother has to manage as best she can. The initial signs of pregnancy toxemia can go unnoticed unless the goat keeper is alert; a slight unwillingness to get up in the morning or go out is often the first indication. This is followed by extreme lethargy to the point where the doe cannot rise at all. The more deficient she is, the earlier the toxemia starts.

Goats that have been on good remineralized pastures and have been properly fed through pregnancy with all their minerals and normal feed should not suffer from this complaint even when there are multiples *in utero*.

Regrettably there are still some people who think that a pregnant goat, if dry, does not need feeding. This is *not* true (nor is it true of humans). CAE-positive does, or those of very advanced years, may develop pregnancy toxemia even when properly fed — although my 10-year-old-plus does never did so. Another contributory factor is lack of exercise. It is most important that pregnant does are sent out to find their feed in the paddock with other does; too much hand feeding can make them lazy.

The immediate treatment used to be drenching the patient daily with a quarter of a pint of glycerine, which the goats loathe, and it was not very successful. But one day I received a call from a friend who was trying to nurse two very good CAE-positive does through to kidding. They had both developed pregnancy toxemia and he had no glycerine on hand. Knowing it was basically due to a mineral deficiency, I suggested he try a drench made from seaweed meal, two tablespoons in water for each animal. He called back to say the recovery was the quickest he had ever seen in spite of the fact that one doe had been very low indeed. In later cases I suggested giving a dessertspoon (2.4 teaspoons) of Vitec stock drench which is much easier to administer than seaweed meal and water. The results are far superior to the old glycerine which kept the pregnancy toxemia cases alive, but did nothing to restore the lacking minerals.

Once a doe has had pregnancy toxemia it would be wise to give her the Vitec on a regular basis until she kids at a rate of five ml every two or three days. It is quite easy to give — use an injection syringe without the needle straight into the mouth. VAM injections would be another option as well in desperate cases.

Prolapse of the Uterus

This is caused by poor muscle tone; it usually occurs in the last week or two of pregnancy and generally with multiple births. Once I had learned to feed my goats their required minerals and get the land in good heart I never saw it again. Once again, prevention is always better than a cure.

The prolapse, which looks like a bag of liquid protruding from the vulva, is really part of the placenta. It is usually visible when the doe is lying down or standing on her hind legs against a fence — usually it goes back in when she is standing up except in very serious cases. When this happens there is a real danger of the placenta rupturing so the amniotic fluid escapes, which means the kids and the doe will possibly die. A dry birth without the fluid required for lubrication can be fatal and artificial lubrication is not always successful.

The best treatment is to buy several containers of the Biochemic Tissue Salt and calcium fluoride (*not* the same as sodium fluoride), which is obtainable from any good health shop. Give the doe three tablets every hour. The tablets can be crushed or she may like chewing them up as they are quite palatable. In the case where I first suggested this remedy the prolapse had been evident for about a week — it cleared up within 24 hours of the treatment. Continue to give the doe ten tablets a day until the kids are born. Calcium fluoride improves muscle tone and could help ensure a normal birth.

Treatment used to consist of strapping a contraption that looked like a coat hanger across the hips with the handle part inserted into the vagina to hold in the placenta. Every time it was used on my does, it ruptured the membranes in advance and a dry birth with a dead doe and kids was the result.

Another suggestion is to insert a couple of stitches in the vulva to hold it shut. I feel this is unacceptable because of the risk of septicemia, *and* the possibility of not being there in time to cut the stitches at birthing time could mean a badly torn doe. I know this is almost universally done in the horse breeding industry and it says nothing for the horse breeders that it should be necessary.

The administration of calcium fluoride has been highly successful every time it has been tried and should be the only cure used — it is completely harmless. Again prevention is better than a cure.

Reactions to Drugs

People frequently have access to drugs of whose side effects they have little knowledge. Their administration is better left to the veterinary professionals who are trained in their use and understand any untoward results they can produce. All drugs, by their very action, have some side effects. Penicillin, for example, causes many animals to lose their appetites by damaging the villi in the intestines.

I was taught by the vets at the University of Melbourne to use a vitamin B12 injection (two cc) with any drug that was administered because it minimized the side effects. It makes the goat feel better which is, after all, the object of the exercise.

Originally goats and sheep were assumed to be equal in their requirements for antibiotics. Unfortunately it was discovered, too late in some cases, that goats could often only tolerate a fraction of the normal sheep dose. In one case, tetracycline drugs (administered by veterinary practitioners) in doses that would be reasonable for sheep, caused irreversible anemia, bone marrow damage and renal failure in goats. This was mentioned earlier in the book. Hopefully it could never happen today, but early experiments with the Ivermectin (a vermifuge) group showed again that goats could only tolerate a fraction of the amount given to sheep.

So make sure that only a member of the veterinary profession gives antibiotics. Do not get "something that would help" from a friend and administer it yourself.

Cortisone

This is normally manufactured in the adrenal glands. It needs two vitamins to be active, vitamins C and B5 (found in barley). Extra doses of vitamin C will help activate the synthesis. Sudden stress of any kind, nutritional or environmental, can deplete cortisone very fast which is why reaching for the vitamin C in an emergency often has such striking results.

However, in humans, the administration of artificial cortisone can stop the natural synthesis for up to two years and there is no reason to suppose that something similar does not occur in the animal world (it should not, because animals make their own vitamin C and humans cannot do this). But it is wiser to stick to the vitamin C for your goats.

Hormones

Hormones are used in synchronization of estrus, ovum transplant, embryo transfer and so forth. All appear to have the same side effects, which are counterproductive to the result intended. They deplete the body of, or interfere with, the synthesis of vitamins A and D and the assimilation of calcium and magnesium.

I have had many inquiries from farmers who have had their goats on an artificial breeding program and have found that when they wanted to revert to normal breeding nothing happened. Goats usually fail to conceive because of a shortage of vitamins A and D. Either administering cod liver oil orally or injecting vitamins A, D and E will start the breeding program up again. They need a teaspoon (five mls) of cod liver oil per week.

I bought in a batch of milkers who had been used for ovum transplants and they had also had their estrus cycles synchronized for easier handling. It took a whole year of ongoing supplementation with vitamins A and D before those does caught up with their own manufacture of the vitamin from their feed. During that time, my own does did not require any supplementary vitamins A and D. I was farming in Gippsland, Australia, the farm was organic, and the correct amounts of vitamins (and minerals) were naturally available from the paddock.

So, if the farm strategy includes using any of the above procedures, and often it is the only way to save or acquire valuable genetic material, make ongoing vitamin A and D supplementation part of the program. The earlier dosing with cod liver oil (a teaspoon per head per week as a routine dose is plenty) is started, the less lasting the side effects of hormones since oil-based vitamins are stored in the liver.

Note: When using intra-vaginal sponges, there is a withdrawal period and the milk should not be used.

B.T.Z. (Butazolidin, Bute)

This would be unlikely to be used for goats as it is generally administered to mask pain. I learned about it from the senior teaching vet at the University of Melbourne in the 1970s who tried it on a buck that had (unknown to us at the time) septic arthritis. He told me that it was a dangerous drug, with the nasty side effect of causing internal and often fatal hemorrhages. We tried it on the crippled animal as a last resort. The result was pitiable, the emaciated animal suddenly found a burst of febrile energy and careened around the paddocks to the point of exhaustion. It was very hastily put out of its misery, after which we found out the real cause of its malady. However, as I write this there has been at least one attempt to ban the use of this drug because it was found in export beef, as reported in a printout from the Shepparton Veterinary Clinic.

Nowadays a great many vets can use homeopathic arnica (see herbal section) which is a totally safe and very effective painkiller. It is a good idea to have a supply of 200c pilule or drops on hand at all times. These can be obtained at health shops or from vets willing to use homeopathic methods.

Retained Afterbirth

This is a rare complaint and should not trouble the goat keeper whose animals are in good order. A lack of selenium and potassium due to chemically fertilized pasture would be the main cause. Cider vinegar, the right minerals in the feed and organically grown pasture are the best preventatives. See the section on sulfur and its role in selenium assimilation.

Do not assume that because you have not seen the afterbirth that it has not been passed because the doe generally eats it — and it is not so very serious if it is retained. However, if you suspect that the afterbirth has been retained, give the doe a dessertspoon (2.4 teaspoons) of vitamin C daily for a week to ten days which will prevent septicemia and the afterbirth should be reabsorbed safely. Should the doe show signs of discomfort, put her back on the vitamin C, which can be given by injection, but the oral route is less painful.

A doe that I had just acquired, and whose kidding date was uncertain, produced two kids simultaneously in the early hours of the day. I found her the next morning with two dead kids hanging out. I removed them carefully; the cervix was closing rapidly so I realized there would be no afterbirth. I gave her 10 cc of vitamin C by injection for the next two days and then two more days of oral vitamin C. She seemed quite well, but ten days later she was definitely off color. There was no smell or discharge, but I felt it was the afterbirth, so put her on a course of vitamin C for a week. There was no more trouble and she kidded next time quite normally.

Scabby Mouth (Orf)

This is similar in appearance to the disease known as Orf in the United Kingdom where it is considered a notifiable and incurable ailment. It is a herpes-linked organism that likes a copper-deficient host and is very contagious for goats at risk through copper deficiency.

The goat will hang back at feeding time, obviously in pain from scabs building up around the mouth. If allowed to develop unchecked, the scabs will cover the face up to the eyes so no hair is visible and will run a term of three weeks.

This disease is quite unnecessary. Make up a mixture in a small bucket as follows: one dessertspoon (2.4 teaspoons or 9 gr) of copper sulfate, the same of vinegar and fill the bucket with water. Dip the goats face in it so the scabby area is thoroughly wet. The scabs will dry up and drop off after two or three applications, sometimes sooner. The diet should be amended to see that all the goats copper levels are correct, this illness will not strike if they are properly fed.

Scald

See foot rot.

Scouring

See diarrhea.

Scurf

See dandruff.

Skin Cancer

Fortunately this is not as common as it used to be. Breeders of white goats realized that they had to breed for tan skins in a hot country like Australia — in the countries of origin, where the sun was not so fierce, the goats had pink or white skins. The tan skin means the goats do not contract skin cancer, while the pale and pink-skinned individuals are at great risk, especially on the udder, nose and around the eyes. Pale-skinned Saanens will crop up from time to time, as the original imports from both England and later New Zealand often had pink or white skins.

Skin cancer starts as a roughening of the skin, turning into scabby lumps which are definitely painful in bad cases, especially at milking time. The milk itself appears to remain unaffected, but the discomfort caused by the cancer would mean ultimately the goat would have to be put down.

The use of vitamin H, PABA, (see section on vitamins), 250 mg crushed in the food daily, could help when the sun is at its strongest in the summer months. In severe cases double that quantity could be used. Vitamins A and D should also be given in the form of a dessertspoon (2.5 teaspoons) of cod liver oil weekly with a dessertspoon (9 grams) of vitamin C daily — both these vitamins are helpful against cancers.

Any doe with a pale skin, whom the sun affects, should always be mated with the darkest skinned buck possible, even if it means going out of her breed. A Saanen crossed with a British Alpine generally produces a dark-skinned kid, almost black in some cases.

Snakebite

Snakebite either kills instantaneously by immediate nervous paralysis or, more usually, by slow loss of muscle control which allows time to deal with the problem.

The eye muscle is the first to relax; the pupil appears to be spread right across the eye. People often call me and say that their animal is ill and the eye "looks all funny and black."

In the section on milk fever I mentioned that the signs were almost identical. Loss of motor control is the next step, followed by death in bad cases, or a long illness if the bite was low in venom.

Give 15 cc of vitamin C by injection intramuscularly in the side of the neck and repeat in two hours if necessary, although often the first dose is enough. There is no use in looking for a vein to do an intravenous injection because when an animal is in a state of shock, as in snakebite, the veins collapse and cannot be found. Failing injectable vitamin C, give a heaped teaspoon by mouth every half hour until the goat looks better. The first time I cured a goat of snakebite was before vitamin C injections became obtainable. The goat was bitten on the mouth. (I did not find this out until two days later). Somehow some of the venom must have landed in his eyes because they had clouded over, so I gave him vitamins A and D as well as the heaped teaspoon of vitamin C. He was staggering a little, but was quite alright half an hour later. I repeated the dose once more. The puncture marks, when they did show up, were on the top lip and it appeared slightly swollen, so I squeezed out some clear-colored fluid.

Keep the patient quiet and comfortable until it is back on its feet and eating well. The great advantage of using vitamin C — pioneered by Dr. Fred Klenner in the 1930s and much used by a Californian dog vet (Dr. Bellfield, DVM) — is that the type of snake is totally immaterial, which is not the case if antivenin is to be used. So often one never sees the snake anyway and vitamin C is also cheaper and more easily available. In my experience (vets tell me I am unlucky), anaphylactic shock to a lesser or greater degree can follow the use of antivenin and it is almost worse than the bite. Another disadvantage of antivenin is that if it has to be used twice in a short time, a reaction is inevitable and could kill.

If the location of the bite can be found — do not waste time looking for it until *after* the vitamin C treatment has been implemented — rub some sodium ascorbate powder well into it as this effectively stops the pain which can be considerable. (I rate a red-backed spider bite as the most painful bite I have experienced, the pain went away within three minutes of rubbing the vitamin C well in). However, often it is not possible, as in the case above, to see the bite marks until the hair falls way from around them.

Goats bitten on the udder are a different story, nothing seems to help the udder. The bite generally does not affect the rest of the animal, but in the one case I had it totally wrecked the udder. Try large doses of vitamin C with extra dolomite, it might work or, as in blackleg, putting the vitamin C straight into the udder might work. I saved the udder on another doe that had been bitten by using hydrogen peroxide as for black mastitis.

Prevention is always easier than a cure — put bells on the collars of the goats. Anyone handy at brazing can make them from copper or brass pipe. Snakes are reputedly deaf, but they can definitely sense the vibrations from bells. I never had another goat bitten once I fitted them and tiger snakes were endemic on that farm.

The vitamin C dosage should be 2 mls to a gram, nothing less. There have been cases of snakebite where the animals died from too low a dose of vitamin C.

Spermiostasis

See hereditary defects.

Split and Peeling Horns

I had two inquiries about this in 1990, it was the first time I had heard of it. It appears that it is due to a calcium/magnesium deficiency and, in both cases, once the goats (fiber) were given supplementary dolomite the trouble ceased.

Sunburn

This has already been covered in the section on skin cancer, be especially careful not to buy a pale-skinned white goat.

Tapeworm

See worms.

Teats (Deformed)
See kid section in Chapter 8.

Tetanus
This illness is due to *Clostridium tetani*, which thrives in deep (usually) airless wounds that have not bled profusely and/or not been thoroughly disinfected. Unfortunately quite often the wound that causes tetanus is not even seen. Gunshot pellets are likely to start it up as are any bullets. Tetanus takes about ten days, occasionally longer, to incubate.

Signs are stiffening and lack of coordination, the classic locked jaw of tetanus (the old name for the disease) and if the chin of the victim is tapped sharply the eyes will roll up. Fever accompanied by high temperature follows and any sudden noise causes the patient to convulse — bad in a goat and devastating in a horse. It is an intensely painful illness and it is better not to allow it to take hold.

Fortunately vitamin C works very quickly to detoxify the clostridia before tetanus reaches the bad stage and should be used immediately at a rate of at least 20 grams by injection with a follow up every hour until the patient relaxes. The longer the disease is allowed to develop, the more pain for the goat and the longer it takes to cure. Both the cases I have treated relaxed within ten minutes of the vitamin C injection, although I gave three more the next day to be certain.

Tetanus immunizations may stop the disease from developing, but they do not stop it from occurring if conditions are right for it. They merely give the farmer a sense of false security — which means that all wounds are not automatically disinfected. The organism is generally found in any soil where animals have been kept (like blackleg on sheep farms, another clostridial condition). Adult animals generally develop a natural immunity and it is more often the younger animals that succumb (except in the case of gunshot).

There is some mention that the tetanus vaccination is not as good as it once was, possibly the strain has mutated, but it is still one of the vaccinations which might help if one only has a few goats. Should an unimmunized goat sustain a wound that could possibly lead to tetanus, it would be a good idea to take it to the vet for a tetanus antitoxin injection. It must *not* have the toxoid, which is the immunization, until at least six weeks after the wound. If you cannot get the vet, put the goat on a maintenance dose of vitamin C at a rate of 10 cc by intramuscular injection the first day and one heaped teaspoon daily for at least 10 to 12 days after the wound, even if it has healed. This should prevent any infection arising.

Tetany — Grass, Lactation, Travel

All tetanies are the result of a magnesium (and possibly calcium) deficiency. Owners of a few goats, who look after them well, are unlikely to have trouble with the first two types. Large goat farms who make sure their goats are regularly supplemented with dolomite and the other minerals are also unlikely to have it.

Lactation Tetany

This is similar to milk fever but occurs further into the lactation. It is caused by a lack of magnesium in the pastures and/or feed, either inherently or because of the artificial fertilizers currently used to grow the feed. On conventional, non-organic farms, a sudden spring growth seems to make the imbalance worse. The victim, which is a lactating doe, suffers from a lack of magnesium which causes a collapse with symptoms similar to milk fever. Once the goat is down, it will struggle until it dies. The goat should be treated with magnesium and calcium; the injection is obtainable from any fodder store. Prevention, in the form of correct mineral supplementation, is, as always, easier for the farmer (and less painful for the goat).

Grass Tetany

This is similar to lactation tetany but happens any time there is a sudden flush of magnesium-deficient grass — often after the autumn break. Work done by Dr. George Miller when he worked in the Gippsland Department of Agriculture at Warragul found that new grass contained nothing of value for the first six weeks of growth. Grass in paddocks that have been top dressed with dolomite does not cause this condition.

I once saw ten cows dead in a circle where they slept after eating the first flush of self-seeded barley. This had grown after a long drought followed by five inches of rain in a paddock that had been regularly "supered." It was also the first time I heard a vet actually mention superphosphate poisoning as such.

The remedy is the same as for lactation tetany — if the animals are found in time. Make sure the offending paddock is analyzed and remineralized.

Travel Tetany

This is when goats either die or become very ill from the stress of travelling. Stress uses up available magnesium very fast. If the animals are not comfortable travelling and their magnesium levels are low, disaster can ensue — dead goats arriving at shows. A teaspoon of vitamin C, and some vitamin B6 as for travel sickness the night before will help them to withstand the stress of a journey — regular supplementation with dolomite should take care of the rest.

If the animal is found before it dies, it will be shaking and in distress. Crush four magnesium orotate tablets and get them into its mouth somehow. If you cannot get them into the goat's mouth, insert them into the rectum. The magnesium is absorbed through the membranes of the mouth or intestine. I have used this with a badly affected horse, it took four-and-one-half minutes for the horse to recover. Dr. Kristin Marriott, B.V. Sc. told me she had used it successfully with cows.

Tick Bite

After a tick bite, the goat becomes lethargic and eventually goes into a coma. If carefully examined, a tick(s) will be found — a dab of tea tree oil applied to the tick works wonders — the tick dies and lets go. Give the goat five grams of vitamin C by injection, it is not too late even if it is in a coma. If only oral C is obtainable, give three teaspoons in liquid slowly down the throat. Take great care if the animal is comatose that the liquid does not go down the windpipe.

There is apparently an antivenin, but it is not effective once the animal is in a coma. Friends with a tick-bitten dog, already comatose, were told by the vet it was too late. Only then did they remember the oral vitamin C I told them about. They dribbled it down the dog's throat and it recovered very quickly.

Animals in tick areas eventually build up an immunity. It might be worth raising the goats' sulfur intake to an extra full teaspoon and a half daily so they would be less likely to be attacked (see section on sulfur).

Tick-infested pastures are always on sour land that needs remineralizing and looking after. Have the land tested if infestations are a problem.

Toxoplasmosis

This disease is carried by cats. It is virtually impossible to control their movements and there is really no way of completely avoiding the occasional outbreak. Toxoplasmosis causes abortion or stillbirth. First kidders are most often affected as older goats will usually have built up an immunity to attack. There are other reasons for abortions and stillbirths and the fetus or dead kid should always be taken to a vet to find out the cause of death — I am always relieved when it is toxoplasmosis because then I know that the particular goat will become immune.

The biggest problem with the disease is that it is a zoonose that can be very easily contracted by a human handling the dead kid or fetus carelessly. It can cause pregnant humans to abort — correct hygiene should *always* be observed when handling sick or dead animals of any kind. If pregnant take extra precautions.

Tuberculosis

This disease is rare in goats. It used to be claimed that they were immune to it, however there is no evidence one way or the other, so care should be exercised. They should not be allowed to have contact with infected cattle — there are still a few about — it is not worth taking the risk.

Signs are similar to other wasting diseases. In cattle it is accompanied by runny noses and coughing. If tuberculosis is suspected, have a vet do tests immediately and on no account use the milk.

Tuberculosis is another of those diseases that strike deficient animals. Goats whose mineral needs are fully met and have ad lib seaweed meal should not contract it anyway.

Udder, Disease of

See flag, hard udder, mastitis and edema.

Unthrifty Kids

See Chapter 7.

Urinary Calculi (Water Belly)

Water belly is painful, but not too serious in a female as the ureter is shorter and wider; but in a male it is painful and often fatal as the ureter is long and narrow. The stones form in the kidneys from unassimilated minerals which are passing down the ureter — or trying to. Sometimes a small stone is passed — a very painful proceeding — but otherwise take the affected animal, generally a buck, straight to the vet. Giving a drench of cider vinegar may help, it will certainly prevent a recurrence. Two teaspoons of vitamin C can also be given daily for two or three days and it may help dissolve the stones and flush the bits out.

This is definitely one complaint where prevention is better than a cure. A buck, even on highly mineralized bore water, will not be troubled with calculi — as little as a teaspoon of cider vinegar per day in the feed is all that is needed. I kept up to seven bucks for many years with only a mineral bore for water. They all received cider vinegar and there were no problems with calculi.

Warts

The virus that causes warts likes a host deficient in magnesium and vitamin C. However, goats, like most animals, synthesize their own vitamin C so only extra magnesium will be needed. Properly fed goats on the right minerals will not contract warts. There is enough magnesium in dolomite to make the warts drop off in a few days. Animals on regular dolomite do not develop warts. If they have them coming off magnesium-deficient land, the warts soon dry up and drop off once the dolomite is run through the feed. Epsom salt can also be used externally as a wash if desired, but it is not really necessary.

Worms

Drenches

In *Australian Goat Husbandry*, written in 1978, I wrote that good husbandry and not drenches was the long term answer to worms. Twenty years later it is truer than ever; drench resistance — which means that the worms mutate to cope with each drench as it is invented — is a fact of life.

When I was doing a talk for the local Department of Agriculture, the convener said as he introduced me: "Well, I hope you have an answer to worms, they are becoming resistant to drenches faster than they can make the new ones." Quite so. Before that, when talking to a Department of Agriculture vet who was monitoring the bloods sent in for CAE testing, he asked me what I was up to now — a fairly common question in the profession. I told him I had forsaken chemical drenches and was using copper and that it was working, his response was, "Thank God for that because soon we will need something that will work and go on working." Again, quite so.

Natural Resistance

As in other species, there are hereditary lines of goats with resistance to worms. I was fortunate to have one such line. They needed one third the amount of drenches that the others did — before I learned to improve my management and keep the worm problem at bay. Careful record keeping will show up the characteristic. However we cannot pin our faith in a number which is probably .00001 of the total population.

Seasonal Upsurges in Parasitic Worms

Strict adherence to the practice and principles of organic farming are really the only long-term defense. Nature did not intend animals to be wiped out by worms (nor did she intend Australia to be inhabited by our domestic stock). David Mackenzie's statement that worms are needed to preserve a balance in the gut at times of extra high protein in the spring pasture was sound in the United Kingdom, Europe and maybe New Zealand — unfortunately not here. In Australia, due to our inherently poor soil, the herbage does not become too rich. Farmers who do not realize that at kidding time (spring) there is a natural upsurge of worms in the gut, will find themselves with dead goats unless the supplements are being fed. Since I learned to see that all goats got their copper regularly in the ration, I have not had to give what we called the "kidding drench." The copper in the system stops a blow-out in the worm population.

Restoring Soil Health to Encourage Soil Fauna

If the soil has been analyzed and remineralized and the pH is in reasonable balance — between 5.0 and 6.5 — earthworms, dung beetles and the soil mycorrhiza will, between them, take down and process the dung just as fast as they can obtain it. Parasitic worms can do no harm underground, they need damp pasture so that the larvae can crawl up the grass and be ingested.

A word of warning here; according to *Acres U.S.A.*, tests running for two years from 1988 in the United States reported that manure from animals treated with the Ivermectin group of drugs was not processed by soil fauna.

Good Husbandry

Good husbandry is the other weapon of the farmer. A goat given a choice will not go out and graze damp grass — they are browsers by nature and worms do not live in trees. On damp days goats must have hay ad lib so they are not forced by hunger to graze dewy worm-infested herbage. In the higher rainfall areas, they should have hay on demand at all times. Overstocking is another potent cause of worm problems.

Worms in Winter and Summer

In areas of Europe and the United States where there is a winter freeze up, and in the drier belts of Australia and the United States, there is virtually a closed season for worms. They cannot operate in freezing or extremely dry conditions. But in the more temperate parts of the world usually chosen for goat dairying (partly due to the proximity of markets) this does not apply — worms thrive all the year round.

Goats are clever animals. Watch them when they go out to graze. If they have a choice and plenty of room, they will first graze the areas that the sun has been on longest, ensuring that the grass they graze is dry and relatively clear of worm larvae and eggs.

Alternating Paddocks and Giving Goats a Choice

Another weapon against worms is to run stock that do not share the same type of worms. With goats, horses are the only animal that is suitable. If the horses are well mannered and handled there should be no problem, but on no account allow the goats to be chased. Buck paddocks should be alternated and spelled regularly and be big enough to put a horse in when the bucks come out, so that the horse can graze them to the ground. Then a few buckets of dolomite can be spread and the runs left until the fresh new growth comes away. I found this a very successful strategy for years and the bucks were hardly ever wormed. Do not run goats with sheep or cattle if it is avoidable. As a preference, allow a year to elapse before running goats on sheep country. If the land has been tested and remineralized, followed by aeration, this time can be cut considerably.

I have always let the goats have the run of the farm — I have "goat bars" on the gates. This is a bar about three feet long with a hole at each end with a split link and strong snap clip in it. This bar is fastened at one end to the gate and at the other end to the fence, keeping the gate open, but horses are prevented from pushing through. Similarly the fences are usually made with one high "sight" (white) wire, with two or three plain strands underneath, confining the horses but allowing the goats to pass where they will.

Signs of Worm Infestation

Signs of wormy goats are runny eyes, picky appetite, lowered milk yield, scouring, anemia with some types of worms and occasionally bottle jaw (illustrated in the section on liver fluke). Any or all of these signs can mean worm infestation.

Copper and Worms

Hungerford, in his book *Diseases of Livestock*, avers that unexplained scouring is nearly always caused by a copper deficiency. After two years of experimenting with copper levels in stock, I would enlarge on that and say that any animal receiving its correct amount of copper will *not* be troubled by worms (*The Albrecht Papers* confirm this.)

It seems that no one really knows what the correct copper requirements are for a goat. Following the publication of an article I sent to the United Kingdom about colored goats and copper needs, some vets decided to work out what these were. Initially they started by trying to fix a fatal level for the mineral. They administered what they considered to be a lethal dose to a white goat (which does not need so much copper as a colored one) and then waited for it to die. At the time of the letter telling me of the experiment (a few weeks after the experiment), the goat had never looked better and flatly refused even to be ill.

Old-fashioned Drenches

The two drenches most commonly in use before proprietary drenches became the norm were copper sulfate with either nicotine sulfate or lead arsenate — both the latter are poisonous and, luckily, unobtainable now — so not surprisingly this mixture killed a few animals. I did know a few people who kept their goats healthy by giving them a plug of good quality pipe tobacco occasionally.

Apparently, not very many people used straight copper sulfate although, in early 1960 I asked the landlords for a worm dose for an old reprobate who passed as a goat (which they had given me), they sent down a tablespoon of copper sulfate. I drenched her with it — not knowing any better or worse. After the copper sulfate the goat looked better and that was that.

Using Copper Sulfate as a Vermifuge

Dr. William A. Albrecht, a highly qualified and much respected soil scientist who studied in the United States, but lectured all over the world (including Australia), did much research with minerals and plants and animals. His work on copper is of particular interest. He found that the Bordeaux mixture we used so successfully on our orchard trees (made up of lime and copper sulfate) did not actually kill the fungus by contact as was supposed. The tree absorbed the copper, and fungus will not stay on a plant with adequate copper in its tissues. Similarly, he found that when animals which had been given copper sulfate recovered from worms, the copper did not actually kill the worms, rather the copper was absorbed by the animal and no worms of any kind would stay in or suck blood from an animal that has plenty of copper in its tissues. I and others had experimented along these lines for 10 years with horses, cattle, sheep and goats, with 100 percent success.

Initially I tried a maintenance dose with the goats which I kept up for nine months, but it was so high that it seemed unrealistic. At that rate (a small teaspoon of copper sulfate per head per day), each goat was receiving the equivalent of nearly two pounds of copper sulfate per year, so I cut the amount to half. After that I did occasionally have to give extra copper. During that nine months, the goats were mated and kidded without having to be wormed as usual. Additionally, the first kidders did not get the almost mandatory dose of cowpox, there were no foot problems in spite of the fact that it was an abnormally wet winter, and they never looked better. Ninety-five percent of them were British Alpines, the other five percent were either Saanens or all blacks, but they all received the same dose of copper sulfate per head daily, run through their feed. The farm was also low in copper.

Permanent Supplementation

The lick described in Chapter 6 including copper seems to work very well with paddock animals. For hand-fed goats I run the copper sulfate through the feed at a rate of a full teaspoon per head per week. Dark-colored goats may need more than this and the rate can be increased; those who run the all blacks should make a note. If the copper is incorporated in the water that soaks the barley (as well as the cider vinegar and any other minerals such as cobalt or boron (borax) that are needed), it is mixed in with the dry feed and also dampens it; so the copper intake is as near natural as can be managed.

Drenching

If a drench is needed — and I do not advocate it because I find it is unnecessary even at kidding time if the normal supplementation of copper has been ongoing — I use a teaspoon of dolomite, half a teaspoon of copper sulfate and a teaspoon of vitamin C powder. Put this dry straight into the mouth from a film container.

Veterinary Reactions

See the comments in the Drenching section (pages 284-285). Several other vets (but not all) with whom I have discussed these worm strategies are interested in the possibilities; they realize only too well we are somewhere near the end of the line with chemical drenches. It is not necessary to withhold milking after drenching with copper. With proprietary drenches, one goat through the milking line by mistake and the whole day's milk has to be thrown out — it's happened to me.

A British *Veterinary Codex* lent to me for notes by Dr. Greg Morrison (a retired veterinary surgeon) did list copper as a vermifuge and gave amounts slightly below what we use now. But, as mentioned, the copper was mixed with either lead arsenate or nicotine sulfate — no wonder the drench got a bad name.

Natural Wormers

Copper has already been discussed at length throughout this book; but there are other herbs and plants that also have an inhibiting action on worms. Honeysuckle and wormwood, both of which goats may help themselves to (preferably through a fence), are two such plants; garlic is not an option where milkers are concerned, nor is it 100 percent successful. Chenopodium oil (oil of American wormseed) was used in Europe for many years by sheep and cattle farmers and may still be a standby. I tried to import it some years ago, but the chemist who was doing the importation was turned down as soon as he stated why we wanted it.

I feel that the last part of this chapter is not necessary, given that the above suggestions mean healthy goats naturally. I include this section only for reference and stress that chemical drenches are *not* the best way to deal with the problem.

Chemical Drenches

In 1990 I wrote that there were roughly 20 drenches on the market; eight years later many of those have been superseded and many more "cocktails" invented. The chemicals, of which there are about 15, remain roughly the same.

A rough rule of thumb for using drenches is *not* to alternate although this was very fashionable some years ago. Vet friends agree that it led to massive, persistant and sometimes insoluble drench resistance problems. Use a drench, be it "white" or "gold," for at least a year, or more if the results are satisfactory, before changing. The final resort, when the goats are dry, is the Ivermectin group under the strict supervision of a vet. It is very powerful and kills everything in the system (including the goat if the wrong amount is used) — both beneficial and otherwise — and the milk from that lactation cannot be used again according to industry printouts when these drenches first came out.

Personally, I used the "white" group — the earliest drench that became available — with satisfactory results for 20 years. But I never "strategically" drenched, animals were only drenched when they showed unmistakable signs of infestation. Then the formula was apparently changed and it started to make the goats rather ill, taking three or four days before they came back on their milk again. This was when I started to shop around and finally had to work it out for myself.

Strategic Drenching

Like alternating drenches, this practice has ensured that any drench used will become useless fairly soon. To drench an animal because it is a certain time of year, *without* taking a worm count, borders on lunacy and yet countless people have done it for years, never thinking what trouble they were laying up for themselves.

It is not good to drench pregnant goats and any goats who have copper supplied on a permanent basis do not need worming anyway. The moment the doe kids, the act of parturition in some way (hormonal) triggers off a massive upsurge in the worm population. Since using copper I have not had to do a kidding drench at all.

Doing Worm Counts

If the cost is not too prohibitive, having a worm count done on the goats at fairly regular intervals is a good idea. If the goats were found to be wormy and still looked unthrifty, the farmer would have to look to the mineral levels and the copper sulfate in the feed could be raised to two grams daily for a while — which could be a lot cheaper and probably safer than drenching. Purchasing a small microscope and learning how to use it is another option. I know several fiber and meat breeders who have done their own counts quite successfully for years.

Types of Worms

An ICI printout of 1978 listed 11 types of worms commonly found in goats. Many of them do not seem to raise problems in my experience, so I have listed below the most usual ones that cause trouble.

Barber's Pole Worm (Haemonchus contortus)

Under a microscope this worm looks like the old-fashioned barber's pole — white with blood bands (its host's) running around it. Until the late 1960s no one seems to have been aware of this worm in the southern states, but it has been a dangerous scourge ever since. It is a blood sucking worm which can debilitate a quite healthy looking adult goat with shocking speed and kill kids even faster. Action must be taken at once if it is suspected. It is not a problem in goats who are receiving the correct minerals in their feed or lick offerings.

The worm only becomes a problem with the warm weather and sometimes due to hormonal activity at kidding. In cold weather it encapsulates itself in the gut, doing no harm (nor is it apparently affected by drenches at this stage). Again this does not happen in copper-fed goats. I used to have bets with myself on the first really warm day of spring as to how many telephone calls there would be saying that the caller's goat had suddenly collapsed. I always told them to drench it — and quickly.

Signs of barber's pole infestation are always anemia — examine the membranes under the eye — and the extreme suddenness of the attack. Other signs of worm infestation such as scouring and runny eyes will show up, but acute and sudden anemia is the chief one. I was brought a kid very late one night that was frothing at the mouth and very ill. It was dark and I feared poison of some sort. I managed to keep it alive half the night, but that was all. The next day I took the body into the local Department of Agriculture. Later on that day, the sister was brought around in the same state, but as it was light, I had a quick look at her eyelids, realized what the problem was and we saved her. Barber's pole worm — a rapid and insidious killer — the report from the Department confirmed the next day.

In the first instance, give a worm drench and a quick acting iron tonic (Ironcyclene is good), and at least two cc of vitamin B12 and VAM by injection; give the B12 every hour if necessary. Injected vitamin C will also be found to be helpful, four to five cc for a kid, double for an older goat (intramuscularly). Once the color of the eyelids starts to return to normal, ensure that the patient receives copper sulfate and has access to its ad lib seaweed meal.

Another insidious characteristic of barber's pole worm is that, unlike most other worms, it has a life cycle of 10 to 14 days so if there is a parasite problem involving more than one type of worm, it will be necessary to give the backup drench on the 10th, 14th and 21st days to be absolutely safe.

Brown Stomach Worm (Ostertagia sp.)

Another bloodsucker, but this one embeds itself in the walls of the stomach to do its task, as well as being found in the gut — it can be active all the year round.

This is not so sudden or dangerous as barber's pole, but action should be taken with goats that appear below par and anemic (check that it is not a copper deficiency, see sections on that mineral and iron). A characteristic of treating this worm is that sometimes when drenching for it, the goat will appear to recover and then relapse again the following day. Apparently the drench can only deal with the worms out in the gut; when they are killed, the ones in the stomach wall detach themselves and start to work. So the signs, generally scouring, will persist and the drench must be given two days running. A buck I leased from Western Australia in the 1970s needed four days of drenching before he was alright — he must have had a very heavy infestation. Ostertagia has a three-week life cycle.

Lungworm (Muellaris capillaris, Dirofilaria immitis)

Persistent coughing and below par animals are the usual sign of a lungworm infestation. Some areas are more prone to it than others. One of the "golden" drenches is usually indicated for lungworm; it will kill the mature and young worms. But check before using that it is specific for both kinds of lungworm. There was much trouble some years ago because some drenches were only formulated for Meullaris, not for Dirofilaria and drenching often did no good at all.

The danger of lungworm infestation is that, if there are a lot of worms in the lungs, the drench suddenly killing them gives the goat a lung full of dead worms and these in effect cause mechanical pneumonia. The goat dies from suffocation — it has happened a number of times. If a very heavy infestation is suspected give a drench that is not specific for lungworm first, some of the "white" drenches are usually suitable, so that the worms in the intestines are killed. Apparently this gives the ones in the lungs a chance to move on and there is not as great a chance of damage when the correct lungworm drench is used.

Any goat that has been infected with lungworm often has a slight chronic cough for life due to lung scarring.

Goats are short on lung area, so try to avoid a lungworm epidemic if possible. Lungworm are not so usual in dry areas, but in some of the wetter places they can be a problem. They have a three-week life cycle.

Pin or Threadworms (Nematodes)

These resemble human threadworms with rather similar signs. They are quite often present without the farmer realizing that anything is amiss. If you see a goat doing a lot of tail wagging when she is patently not in season, have a good look — often the worms can be seen crawling round the anus and setting up an irritation.

Piperazine drenches are highly effective against these worms and I never found any of the others on the market had any effect. Piperazine is not specifically mentioned for goats, but appears to be completely safe. The same powder that is used for poultry can be used quite successfully. Pinworms usually come in with a load of hay, particularly pea hay that has been harvested off old sheep pastures. The eggs are scraped up off the ground with the hay after harvest. They are not a problem in goats who receive their copper.

Roundworms (Strongyle)

This is the most common type of worm and these days it does not rate much publicity. It has a three-week life cycle and occasionally, with careful watching, it can be seen in the droppings. Practically any drench will work against them.

Tapeworm (Moniezia expansa)

Goat tapeworm is reputedly species specific and is not the same as those carried by dogs. I was told this was so, but I have heard of transmission that suggested a crossing of species on more that one occasion.

Very pot-bellied kids are quite often infected with tapeworm and careful examination of the feces will sometimes detect the typical white segments of the worm.

It is fairly rare and, if suspected (repeated drenching for other species of worm seems to make no difference), have the vet test a sample of manure and they can then advise on a drenching program. For years there was only one drench for these worms (Mansonil). Kids and young goats are generally affected, adult goats appear to develop an immunity to tapeworm, which lives in the soil on many farms. The intermediate stage is a soil mite — avoid contaminating pasture if possible as it causes unthrifty kids. Tapeworms dislike copper even more than other species of worm.

Hydatid

These are the exception to the species specific label and can be contracted by goats in the same way that sheep (or any other animal, human included) catches them. Dogs and humans can pick them up from rabbit livers or meat. An area subject to hydatid, and anywhere that sheep have been farmed, is at risk. Make sure the goats are treated for hydatid if necessary. Consult your vet.

Summing up

Having rechecked this section and revived memories of many struggles with all the types of worms mentioned, I realize how trouble free and uncomplicated life has been since we learned to use copper sulfate. Animals are healthier without the poison drenches and I am certainly a lot more relaxed.

Wry Face, Wry Tail

See hereditary defects, Chapter 11 and Chapter 8.

Medicine Chest Supplies

It will be useful to have as many of these items on hand as possible. Several of them, such as seaweed products, dolomite and copper sulfate will be part of the feed shed anyway.

Vitamin A, D and E injections
Vitamin A and D using cod liver oil
Aloe vera lotion or gel
Arnica — pilule or drops, 200 C strength

Bandages — wide stretch variety
Calcium fluoride tablets (Biochemic Tissue Salts)
Comfrey — tablets, lotion and cream if available
Cooking oil — for drenches
Copper sulfate
Gamgee tissue — for wound dressings (as used for horses)
Disinfectant, Dettol, six percent peroxide or similar
Dolomite powder, finest possible — works better in drench
Drenching bottle — smooth-necked wine bottle
Drenching equipment — for large herds, must be calibrated regularly
Empty pill jars or film canisters — great for administering powders
Flint's Medicated Oils — for wounds
Garlic tablets or capsules
Hydrogen peroxide — six percent pharmaceutical grade
Iron phosphate tablets (Ferrum Phosphate), Biochemic Tissue Salts
Lubricating fluid — for births
Magnesium orotate tablets — usually 400 mg
Methylated spirit, iodine or alcohol — for disinfecting navel cords
Needle and thread (upholstery or surgical) — for stitching
Pestle and mortar — for grinding tablets, or two spoons will work as well
PABA tablets — for sunburn
Pharmaceutical chalk — for poisons, dolomite will do as well
Plastic gloves — elbow length for kidding
Purple spray, if obtainable — comes and goes on the market
Salt — for disinfecting
Scalpel or sharp, pointed knife
Seaweed meal (urea-free)

Septicide — excellent fly repellent and disinfectant ointment
Slippery elm powder
Sodium ascorbate powder (vitamin C) — tasteless and can be mixed with milk
Splints — ladder type from vet or homemade
Sulfur powder — not to be confused with sulfa drugs which should not be used
Syringe needles, sizes 18 and 21
Syringes — large for drenching: 60 ml, and small: five-, ten- and twenty ml sizes
Tea tree oil
VAM — Vitamin, mineral and amino acid injection
Vitamin C injections, Sodium ascorbate — for intravenous and intramuscular injections, ascorbic acid for intramuscular only
Vitamin B1 injection
Vitamin B12 injection
Vitamin B15 DADA injection — good for the liver, obtainable from a vet
Vitamin B5 tablets
Vitamin B6 tablets
Vitec stock drench — a seaweed drench, with no extras, chelated or otherwise
Worm drenches if used (*not* recommended)

Chapter 12

Breeding & Selection for Desirable Characteristics

Whether the farm is commercial milking, fiber, meat, hobby goats kept for show, family milk or pleasure, the aim is the same — to breed a better animal for the job at hand.

Objects of Breeding

There are specific traits to concentrate on, good let down of milk, length of lactation, evenness of lactation, the ability to be good foragers, good fleece type, udder type, temperament, hardiness, quick developers (for the meat trade) and so on — everyone has an ideal in mind. The list is endless and each breeder/farmer will have fairly clear-cut ideas on the most desirable characteristics.

Angora breeders know to their cost how easy it is to breed kemp *in*. Ill-advised crossbreeding to increase the number with pure milk types 20 years ago has left a legacy of kemp that many breeders have only just gotten over trying to breed *out*. Angoras upgraded from feral goats with a predisposition to fiber do not suffer from the same problem. The set of the horns often shows if the ferals carry Angora recessives. Now that the Texan and

South African breeds are in Australia, hopefully the kemp will be further reduced.

Boers are being upgraded from ferals, Angoras, Cashmeres and by a few long-sighted people, from Nubians. This last seems to me to be the most logical. Nubians have been known in North and South Africa for a great many years. In fact, I have seen a print of them in the Chinese gold-digging era of the last century in Australia; it is possible that they came from Africa. Boers as a pure breed are reckoned to be about 30 years old and before that the type was pretty well fixed and Nubians had to figure in their development. Certainly those that have used them to upgrade their stock seem to be quite successful.

Long Lactations

In the British Alpine breed long lactations appear to be dominant; if healthy and well looked after, they will milk for years on one lactation. A friend called me the other day and told me that their family milker (only kidded once at two years) had just died at fourteen. In the latter few years her milk had dropped to two pints a day, but she had averaged double that for most of her life. What more could be asked of a family milker?

This trait was peculiar to the Old English goat; it was bred by the peasants of bygone days to milk for a lifetime on one mating — they could neither afford to take their milker to the buck each year nor to be without milk. This goat was bred out as a distinct type about the time Australia was first discovered by the British. Fortunately, or we would not know now what the breed was like, Captain Cook took goats on his voyages of discovery and those goats left nuclei in odd places such as Arapawa Island off the coast of New Zealand. No other colony of these goats in pure form exists today, except those now being hopefully "rescued" in the United Kingdom. When the islands were explored in the 1970s, several goats were caught and brought back to New Zealand for evaluation. Researchers compared their skeletal conformation with those milk breeds known in the Antipodes today. They found that there were no differences except in the pelvis, and the diagrams showed that British Alpines are the only breed with similar bone structure — and apparently the long lactating gene.

However, recently I have discussed this with Dr. Phillip Sponenberg, D.V.M., Ph.D. of Blacksburg, Virginia, who has done quite a bit of research on the subject and concedes the genetic link, but doubts whether those goats actually came from the United Kingdom. It was more usual to pick them up at ports of call, in which case Capetown or the African coast would be a more likely source. Since that is where most of the First Fleet ships landed, he is probably right.

Conformation

If you are looking for an outside buck to use in breeding, be certain to avoid any who carry the faults of your doe. If you are raising a milk breed, be certain to have a good look at his mother and sisters; if you are raising fleece breeds, a micron count (taking his age into consideration) will always be your safest option; meat goats need to be of tough, sound conformation. Look for points that you would rather have, remembering that breeding does not always turn out as expected; still, one has to try. In horse breeding, a mating that produces more than expected is referred to as "nicking," the same applies with goats. Occasionally a quite unlikely mating will produce exceptional progeny, if it does — repeat it. On the other hand, two show winners, quite good in every respect, can produce an absolute disaster. Some bucks throw their genotype, the characteristics of their ancestors, and still others throw their phenotype, producing stock like themselves. Consult Chapter 4 for illustrations.

Heritability

Some traits are more heritable than others and faults are often only too easy to pass on while some of the good points are quite difficult to "fix." If the farmer/breeder concentrates on a particular point — like high, wide udders with good teats, fleeces with a really good micron count, meat goats that flesh up well and evenly — and culls any animal that does not reach the required standard, over the years the herd will become almost prepotent for that particular type of udder or super-fine fleece or early maturity. This seems to be so true that one can then occasionally risk bringing in a doe or buck slightly lacking in those desired respects, but carrying other *wanted* characteristics, and be fairly certain that the faults can be overcome quite quickly.

The easiest way to demonstrate heritability is the inheritance of horns. As mentioned in Chapter 8, the degree of horning is variable within the polled or horned types. This allows one to sometimes breed lightly polled animals together quite safely without expecting infertile goats. The chart below shows a fairly simple explanation of Mendelian inheritance with regard to the horned and polled factor.

Examples of Mendelian Inheritance

H= Horned, P=Polled

Example 1
Buck (H)   Doe (P)
   K1   K2   K3   K4
   (H)   (H)   (P)   (P)

Example 2
   Buck (P)   Doe (P)
   with Dam (H)
   K1   (P)   K3
   (H)   (P) but carries (H)   (P) pure and
                                                 possibly
                                                 infertile

This works correctly for horns until something happens to disprove it. Mendel's law was later shown not to be quite as immutable as he claimed; however, it is reasonably safe when working out some of the faults carried in polled lines (see section on cystic ovaries and spermiostasis).

Early on in my goat-keeping days I took a calculated risk by breeding a polled doe, whose father had mild spermiostasis, with her polled grandson. The doe's father was able to serve two or three does successfully before going completely infertile. The grandson's sire had been polled, but his mother, though polled, was by a horned buck. I was fairly sure that if the produce were multiples, one stood a chance of being infertile. She had three bucks, two polled and one horned. I sat and looked at the polled pair. By Mendelian principles, one could be pure for poll (and possibly infertile), the other carried the horned gene.

One of the polled kids looked faintly effeminate, so I kept him until he was proven as a breeder (I had orders for all three). I suspected he could have spermiostasis and he did. One does not always have such a clear-cut case, but, in theory at any rate, the above reasoning can be used when working out which way a fault will go. One will have it, one will not have it but be carrying it, and the other will not be carrying it. Only further breeding will tell you which one. If there are four, Mendel postulated that two would be carrying it, and the other two would be a clear positive and a clear negative. This all sounds very easy, and the principle can be used when tracing faults back to their originators. A study of a book on genetics will reveal all kinds of imponderables, like crossing over, so it is not quite as easy as it seems, but one can try.

Recessives

Recessives are from a trait that is carried by *both* sides of a breeding pair, but which may not show up for generations. Some years ago a Saanen breeder who had bred her goats pure for years was considerably astonished to breed three very nice looking, true-to-type British Alpines. Research could not pin down their origins because it was too far back, but obviously the color, which had not appeared before, had been carried for a long time. As British Alpines were "made up" in the United Kingdom, and have been crossed with all breeds there in the process, it was probably a foregone conclusion that they would surface occasionally in odd places. Actually, recessives are extremely valuable genetically, they may revive lines that have been long gone — they may also bring up absolute genetic disasters.

Officially British Alpines are a recessive breed (and possibly French Alpines are as well), and have to breed true when two are mated, when they *must* resemble their parents. Unfortunately that has been disproved too, as I have had one white and one black from two British Alpines on two occasions. In each case their parents were low in the appendices and bred up from Saanens.

Officially (again) brown is dominant over black, and white is dominant over both, but it does not always work out that way. Black, either as in British Alpines or all black, is a recessive color that can crop up anywhere (genetically this is an incorrect statement), and frequently does.

However, major faults like one-sided udders, which have plagued all breeds at some time, can often take up to fourteen years of breeding before the culprit can be picked out. Only after that time, when two individuals carrying that goat's blood are eventually mated, does the fault show up again. All too often a buck is dead before his faults are known and he is nearly always dead before his excellence is realized.

Breeding Cycles

In milking goats these are governed by the length of daylight, provided of course the animals are receiving the proper minerals — copper-deficient animals will not cycle at the correct time. For practical purposes we say that goats furthest from the tropics will come into season in the autumn and bear their kids in the spring. The nearer they get to the equator, the longer the breeding cycle extends, until some breeds only have a couple of months when they are not cycling. Nubians, like fiber goats (and possibly the

Boers), quite often come in season in the summer months, even in the southern part of Australia.

Dairy goats appear to be far more rigid than fiber goats and ferals. At least one large commercial goat farm in the United Kingdom uses lights to control estrus so that they have does kidding earlier or later than usual according to how the lights are manipulated. This means that with goats that do not milk through, they can maintain a year-round milk supply. This method does not upset the goats in any way and has none of the unfortunate effects sometimes associated with artificially induced breeding.

One American authority, Dr. Samuel Guss, D.V.M., postulates that wet weather also has a bearing on goats coming into estrus. Certainly in dry countries, like Australia and parts of United States, does often do not start to cycle until the autumn "break" (rains). I personally feel it is more likely to be something to do with the advent of Spring. If the goats come in estrus early, I find we usually have an early Spring.

Remember, in the northern hemisphere Australian spring is in your fall and our autumn in your spring — even after nearly 40 years in Australia I have to think twice occasionally.

Mating

One important fact must be remembered with regard to mating. If a buck should be really sick, when he is better he will be able to cover does successfully for about 10 weeks afterwards. Then, about twelve weeks later, he will go through a period of infertility lasting the length of time of his illness. This is because sperm is stored, and he will have been unable to replenish the supply during the time he was sick. His elevated temperature when sick may have also precluded the sperm being viable. His long-term fertility should not be impaired, however.

The goat world is full of bland generalizations. "It's quite easy to tell when a doe is in season," is one of them. Sometimes it is indeed, the doe bleats (yells would often be nearer the mark), is damp round the tail which does a lot of wagging, and tries to mount other does if she has company. Other does show none of these signs — one old dear of mine used to give the bucks one long considering look as she walked past their pens, and that was all. In the final analysis, the buck is the only one who really knows.

For single goat owners, a buck-impregnated rag (kept in a sealed container when not wanted), hung in her shed will keep a doe in season quiet. She will stand and wag beside it when "in" — this is particularly useful if she is kept in a built-up area where neighbors might complain about the unusual noise.

If the doe is properly in, she will, when introduced to the buck, stand and wag her tail (sometimes they do a dash round if allowed to, but it is better to have her on a line), if not, she will resolutely refuse to have anything to do with him. If she is ready, the buck will sniff around a little, mount and serve her; as the penis enters her vagina, he ejaculates and arches his back — which is how one knows he has successfully covered the doe.

One service is enough. Even so, some owners don't think they have had their money's worth if the buck does not cover the doe two or three times. If he has served her successfully he usually declines to do it again — for a while anyway. Having the doe on a line means she can be removed without upsetting the buck. One learns to judge very well when to shut the (inward opening) gate in his lordship's face.

Temperaments of both bucks and does can change when breeding. Quite calm does become almost lunatic and bucks, soppy to the point of idiocy in the summer, become arrogant and aggressive — especially if another buck is in sight (perhaps they are showing off).

Does may continue in season for a day or two after service. Dairy does come in season about seven times each year. The first cycle is often quite short, but three days may occur further into the winter. By the end, the period on heat becomes shorter and it may be a matter of hours not days. Breeding goats late in the day is reputed to produce more females, but actually the degree of horning is also a factor. A horned (disbudded) buck — regardless of the status of the doe — produces on average 60 percent-plus females and a polled buck 50 percent. Over three years I found this was remarkably accurate although I had doubts when I first heard it.

Pregnancy (and Cloudbursts)

A doe carries her kids for 150 days — easily calculated if one counts five months and subtracts four days. Early births, three days or more, invariably mean the doe is not 100 percent; it can be due to trauma or nutritional stress, usually the latter. Late births, on the other hand, do not always mean buck kids and do not seem to denote anything wrong. A preponderance of buck kids generally means too much alfalfa or goitrogenic feeds at conception and a deficiency of iodine. Bucks born healthy and does dead or dying is a classic sign of iodine deficiency.

If a maiden, see she gets her minerals and a small feed as usual for the first three months, gradually raising her feed until she is receiving the same as the milkers by the time she kids. Does that are milking should be fed as usual whether they dry off or not; this feed before kidding ensures a healthy beginning to the lactation. Dry pregnant does *must* be fed. Keeping an eye on the vulva will tell if the doe is pregnant. By kidding time the vulva should be loose and soft. If, by the fourth month, it looks normal and dry, she may be doing a false pregnancy. She will come out of it around the date she would have kidded; a wet back end and sudden decrease in size is the result. She should be mated as early as possible next time.

Pregnancy tests are now available and should save much time and worry. I have never used them, preferring to trust known signs; ultrasound is the flavor of the month these days, use it sparingly — I know some Alpaca breeders who have felt their animals were disadvantaged by it and certainly several horses that were similarly affected.

There is much we still have to learn about these electrical devices.

False pregnancies are more common in goats whose diet is inadequate and in polled dairy goats, especially if they are noisily in season for a long time on the first estrus of the year and have not been covered at that time. In those cases it is better to give in gracefully and take them to the buck. Cloudbursts upset the does' normal sequence and are a nuisance.

Kidding

Scanning (see above section about ultrasound) is available to see if a doe is indeed pregnant. There are occasions when it is very important to know this, especially with artificial breeding, when the cost has to be justified. For those who have to wait for nature to take its course, there is little way of telling whether the doe is pregnant or not until about the third month. The vulva of a pregnant doe looks soft, but some show up quite early and others only in the last month. In the last two weeks of pregnancy, a hand placed just in front of the udder and lifted will feel limbs and this method may also be used during kidding to feel how many are to come.

As parturition approaches, the doe's vulva looks extremely soft, but about five or six hours before the actual event, it suddenly looks withdrawn. Soon after this, the first stream of clear liquid can be seen. Most does kid very easily and, often while the new goat keeper is going into a spin wondering who to telephone, the kids will be on the ground by the time the owner gets back to the doe. Some does definitely prefer company, while others choose the farthest corner of the paddock. The "water bag" is often, but not invariably, the last stage before feet, noses, or other parts appear.

Abnormal presentations and kidding troubles are rare in healthy stock, but goats raised on poor land, who are potassium deficient, or over-fed and over-fat goats can have trouble. The normal presentation is front feet and head — I imagine this as the kid kneeling in the uterus with its front feet in the cervix — and in good births two or three will follow each other in this position. The legs appear first and the kid's nose will be level with its knees. If the head is turned back, it is a good idea to scrub up (short nails, clean hands and plastic gloves if the farmer's hands are cut or scratched) and pull the head forward. The kid can be born with the head turned back, but it is not so easy.

Ease the kid out as the doe contracts and give it to her to wash and suckle or use whatever system of rearing has been planned. Each kid must have its ration of colostrum, the first thick milk that contains the antibodies for *that* kid. It also has a laxative action which makes the kid pass the first black manure, the meconium (this used to be called beastings). Colostrum only confers immunity to the goat's own offspring, so giving colostrum from one doe to another's kid is a waste of time. There is also the (to me) unacceptable risk of spreading disease (see Chapter 11 on CAE for artificial colostrum).

Warm up the milk slightly by standing the bowl in hot water and offer it to the kid. Stand the kid between your calves and bring the bowl up to its face, *not* the face down to the bowl. It will drink straight away if it is healthy. This is quite a good strategy even if the doe is to suckle the kid, because the kid will remember how to drink from a bowl which may be useful later on should the mother be unable to feed it for any reason.

The main presentations which differ from the norm are rear legs first and breech. If the former occurs, see that the back legs are straight coming into the cervix, if not, push the kid back to straighten them. Do *not* straighten them in the cervix, this can cause damage.

Help the kid out as before, but once the navel cord is out and broken, hurry the process up because the kid will suffocate as it can no longer breathe through its navel cord. Should it have sucked in amniotic fluid or mucus, dangle it upside down to get rid of it or, in an extreme case, twirl it round so centrifugal force clears the passages.

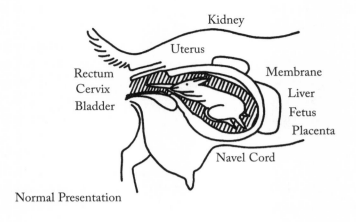

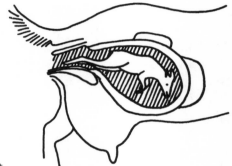

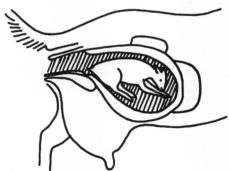

Possible positions of kids prior to birth.

Breech births mean a tail is all that appears (or does not). If the birth is not going according to the rules, investigate as described above and, if the tail is all there is, either push the whole lot back and try to get the back legs up or put the fingers round the tops of the back legs and pull the kid out with them still forward.

When investigating it is sometimes easier with the eyes shut — one thinks through the fingers. Two other possibilities very occasionally happen. A complete absence of action within an hour or so after the mucus has come and the water broken can mean the kid is sitting with its back to the cervix as though watching a TV in the front of the uterus. Patience is needed to slowly turn the kid in a somersault and bring it out by whichever legs become handy first, it will probably have to be delivered upside down, but they are very flexible. I've delivered one kid like this and countless lambs (the latter for other people).

Another catch can be two feet sticking out and nothing at all happening despite either gentle pulls or the mother's contractions. Investigation can show that the legs do not belong to the same kid. Push one back, and deliver them in their turn. Kidding is far easier if the doe is standing; a very experienced goat breeder proved this to me years ago, and even if it means having someone to support the doe, I do it where possible. There is maximum room when the doe is upright. Quite often old does will kid easily in any position, but youngsters sometimes want all the help they can get.

Wherever they kid, disinfecting the navel cord is the first most important task. If this is not done there is a very real risk of infection, especially on small, overstocked farms. Use iodine, methylated spirits or any alcohol if these are not on hand.

The afterbirth should follow soon after the birth of the kids. If you were not present for the entire process, you may have missed seeing the doe eat it, so do not worry too much. This is an atavistic trait from the time when does kidded in the wild and did not want a predator to know that they had done so. The minerals in the placenta are also believed to be valuable to the doe. In Chapter 11 there is a section on retained afterbirth.

Consistently early kids (more than four days early) *always* means there is a health problem; late ones do not seem to matter, but attend to the mineral levels if there is a spate of early births.

Cider Vinegar

The addition of cider vinegar to the diet makes a huge amount of difference; most of the abnormalities shown above just do not occur when does are properly fed. When the blood vessels to the uterus and cervix are operating correctly, the business of keeping the fetus moving and, finally, into the correct position for birth goes on as it should. (For more on the benefits of cider vinegar, see Chapter 10.)

Chapter 13
Goats for Milk

Buying Milkers

Two factors make a good goat farm, the expertise of the farmer and the genetic potential of the goat herd, in that order. It used to be said that environment and heredity ran 50/50; several quite eminent judges and breeders of livestock have said to me, and I agree, that it is management (90 percent) and heredity (10 percent). In other words, it does not matter how good the genetics are, without top-class management, the genetics are wasted.

Goat farms are often started by people who have little knowledge of husbandry in general and even less of goats in particular. They buy any animal available, knowing little of its background — no cow dairyman would take the risk.

If records have been kept, the sellers should be able to tell the farmer something of the breeding and whether the performance of the goats live up to their backgrounds (if the husbandry is good enough). Figures will be available, if the goats have been officially recorded or the milk has been sold regularly, and should show how they perform through the autumn and winter. Goats that start the beginning of the lactation with six and more quarts and then tail off to less than half in a couple of months are completely useless commercially (and probably for any other purpose). They do not even make good family milkers unless one wants to drink frozen milk for half the year.

Udders are all important. Serious consideration should be given to firm, high udders both for hand milking and milking machines — with teats that are easy for both. Very soft-uddered does, no matter how prolific, are not going to stand up to years — possibly not even one year — of machine milking. The animals will break down and end up dragging their udders on the floor. Pendulous udders are to be avoided at all costs (see Chapter 4).

If prospective purchases are dry out of season, find out why — and more importantly how — they were dried off. The latter point is very important. If hormones (stilbestrol for example) were used, it will mean that the goat may not produce well for at least a year and the udder may be damaged. Drying a goat up too suddenly often leaves the udder full of lumps, either with, or prone to mastitis. Beware of very oversized goatlings, they may have been forced and in that case will not make good milkers initially (possibly not ever).

The compact, medium-sized goat is often a more viable option for any concern, commercial or otherwise. They are generally tougher, do not need so much food to be productive, and milk just as well as their larger sisters.

Management

As explained in Chapter 8, it often does not pay either physically or commercially to leave young goats empty for two years. Those that are kidded at 14 months and then milked through the two years (i.e., two years between kiddings) will produce more milk over a lifetime than if they had kidded at 24 months and kidded each year. Goats managed this way often give more milk on the run through than when they were first kidded. The next spring after kidding they will flush up with about three quarters or more of their initial production.

In Chapter 3, different kinds of commercial husbandry are discussed. In the northern hemisphere particularly, many commercial herds are housed all the year round and their fodder is grown and processed on the farm. Others are run on a strip grazing system during the warmer months, especially in Europe, where the weather is, on the whole, more stable than in the United Kingdom.

In 1978 I saw two well laid out commercial dairies in Southern England. Each was milking around 200 does and collecting milk from producers in a forty mile radius. In each case all the milk was processed on the farm into cheese and yogurt and also sold as whole milk. The sales outlets were roughly within the same radius as the collections, so the two operations could be performed together — the refrigerated van was the property of the employer. Each farm bred its own replacements, as well as buying a few top does and/or bucks at intervals to improve the genetic pool. These were bred by specialist breeders and had impressive show and milk records (a doe in the United Kingdom can not be shown unless she gains her milk qualifications at the show).

English commercial dairy.

The goats were fed excellent quality homegrown hay and concentrates, but apparently no silage. English pit silage does not appear to be satisfactory for goats, unlike some of the Australian round bale silage. By 1988 at least one commercial set-up was useing silage, with apparent success.

Cleaning out (always a problem) was done with a tractor and blade. The shed was as depicted, divided into four bays, with a cross-shaped central passage. The goats were drafted out in fifties for milking and, once a week as each lot went out, their area was scraped out and new bedding of old hay or sawdust was put down. The milking parlor was herringbone 10 aside (in which the goats were not fed) and a staff of at least 10 people, including the manager who did much of the driving, was employed.

This sort of option may have investment potential because there is no doubt that the demand for good quality goat products far exceeds the supply. In Australia many of the inquiries for goat's milk products come from other countries, but they are not interested unless there is continuity of supply.

That is the top of the scale, as is the figure that was quoted at a recent goat milk seminar in Victoria of $16,000 to $20,000 to set up any operation. If a farm is rented, either wholly or on a share-farm basis and recycled equipment used, the actual goat operation can and has been set up much more reasonably. A simple, but effective arrangement can be built for about $7,000 to $8,000, assuming of course there are already sheds that can be adapted on the farm. These figures would have to be reviewed.

The cost for such an operation in 1987 was as follows:

Milking machines	$1,000.00 complete
200 gallon vat	1,000.00 complete
New rubbers*	200.00
Platform (wood)	300.00
Cement for yard	800.00
Wood for races, yards and gates	200.00
Paint for vat room	250.00
Electrician	1,000.00

*These figures included air and milk lines, inflations, rings in releaser, claw rubbers, sight bowls, etc.

The machines and vat were acquired at a farm clearing sale, a wooden platform, yards, gates and vat room (the dairy area was already concreted) were homemade. The electrician was the only outside labor used.

The goats in that operation were bail fed while being milked — 10 side by side from the rear. The number of milkers was about 65, with 100 as the top projected figure.

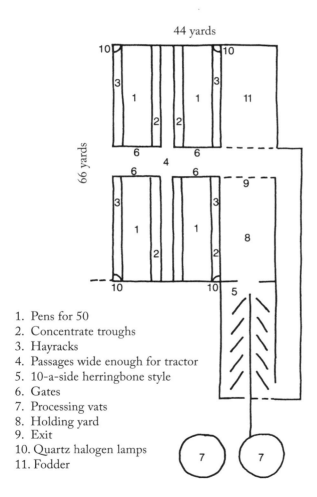

1. Pens for 50
2. Concentrate troughs
3. Hayracks
4. Passages wide enough for tractor
5. 10-a-side herringbone style
6. Gates
7. Processing vats
8. Holding yard
9. Exit
10. Quartz halogen lamps
11. Fodder

English commercial dairy for 200 goats.

This set-up was generally operated by the farmer alone, although he was helped occasionally by his wife. It en-tailed using common sense and doing most of the building work himself and setting up a viable concern at a reasonable figure. The milk was sold to a cheese factory about 30 miles away and was delivered weekly. The growing family of children, the eldest was 11, also gave a hand on occasion.

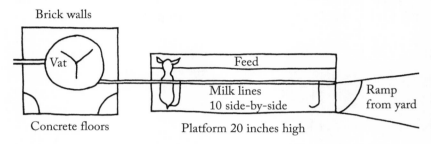

Simple Australian commercial milking operation for 60 to 100 goats.

The price of the goats is not included in any of these breakdowns and is difficult to assess. Supply and demand varies and most importantly the lack of availability of good class, disease-free milking stock is the main problem. The best method is to allow two years to set up while continuing to run the farm as usual, or the owner continues to go out to work. Meanwhile a herd is built up by buying good, clean, young stock at reasonable prices and rearing them. This two years is also a good learning period if the farmer has not handled goats before.

Hand Milking Small Numbers

Hand milking a small, highly productive herd is another option for the single-handed farmer. Papers presented at the Brazil Conference on goats suggested that to milk less than thirty goats by machine was uneconomic both in time and money. (There was mention of one goat farmer in the United States milking 80 goats by hand, but numbers of employees were not available.

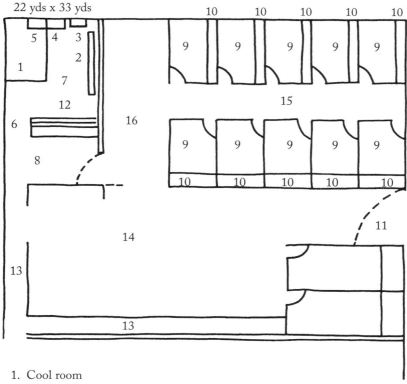

1. Cool room
2. Cooler
3. Water heater
4. Sink
5. Water for cooler
6. Swing door
7. Dairy
8. Milking parlour
9. Stalls 1-3 a piece
10. Feed troughs
11. Gate to farm
12. Bail (see picture to right)
13. Hay racks
14. Lounging barn
15. Passage
16. Route taken by goats

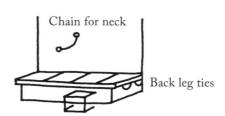

Solid, easy to clean bail, 12 inches high x 88 inches wide. Slabs project 1.5 inches to allow for milker's feet.

Shed for hand milking more than 30 goats.

I ran an operation on this scale for many years and the goats, while not providing a holiday income, kept me and paid the farm expenses, running costs and improvements. I generally milked 25-30 goats. The total number on the farm, counting replacements and bucks, was usually about 45 head.

The milking parlor must be better arranged than it would be for machine milking because of flies. A concrete bail was used and the goats were milked into a stainless steel bucket with a three-quarter lid and the milk run through a six-foot ripple cooler with a filter sock at the outlet. The water run through the cooler was kept as near to 28 degrees F as possible and circulated from a tank in the cool room (the water had glycol added). After experimenting with colors, I found that the fly problem was minimized if all walls and fittings were white (Colorbond, a commercial zincalume, or painted concrete).

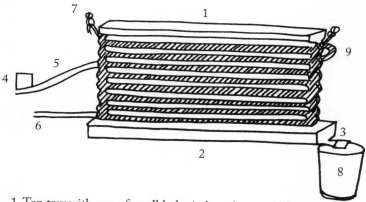

1. Top tray with row of small holes in base (removable)
2. Bottom tray with pipe out (removable)
3. Sock filter on pipe
4. Electric washing machine pulling freezing water from tank in cool room
5. Cold water inlet
6. Pipe back to cool room tank
7. Hanging chains
8. 25-quart bucket (lid keeps out dust)
9. Pipes containing coolant

Ripple cooler.

The milk was stored in five-gallon, food-grade plastic buckets in a cool room kept at 28 degrees. It does not freeze at this temperature when in plastic buckets, but remains at approximately 32 degrees (freezing). The cool room motor had to be slightly more powerful than required owing to frequent, scheduled electricity cuts. During a hot Australian summer, the milk rises one degree an hour when the motor is not working. If the temperature can be brought down to approximately 16 degrees below freezing, according to the time of the scheduled stoppage, not too much damage is done. Electricity stoppages leave the farmer doing some nail biting. This milk was delivered once a week to the same cheese maker as in the operation above. The journey took half an hour during which time temperature rose one degree (in an uninsulated van).

The set up cost was between $5,000 and $6,000 AUD for concrete, cool room, electrical work and plant. The cool room was new, as was the motor originally. The amount of milk delivered weekly was very little short of that from the previous farm that milked twice the number of goats, the extra personal contact and care presumably making the difference.

Free Time

Judging from a paper given at the 1987 Brazil Conference on goats, free time is a worldwide problem in goat milking concerns. Cow dairies generally have a dry period during which the farmer and his family and workers can get away. With goats, seven days a week, 52 weeks a year becomes a chore after a few years, if not before. Various countries have their own ways of dealing with this problem. In Norway, after the first four or five months of the lactation, the milking is cut to once a day — estimated loss in production is about 20 percent. In France, Sunday night milking is generally missed; the farmers admit it takes until about Thursday for the goats to totally settle down again and estimate the loss at about four to five percent. In Australia, relief milkers are preferred, but their availability and cost (at cow dairy rates) makes their employment fairly rare.

Small Milking Sheds

Even sheds or areas for milking one or two goats need to be easy to clean and manage. Goats are great creatures of habit and soon become accustomed to being milked on a proper bail. The easiest to arrange for hand milking is a bench at the side of the shed, it needs to be about 12-inches high and 14-inches wide and made of concrete or cement slabs with no cracks in which milk can collect. In the early days I tried beaten earth, then bricks, then concreting them in, but plain concrete is definitely the easiest to manage. All of the above should be able to be hosed off easily — building in a slight slope helps. The goat is fastened to the wall by a chain which goes back to a loop above her head. A useful adjunct to build in to the bench are two rings to which each back leg can be tied at break-

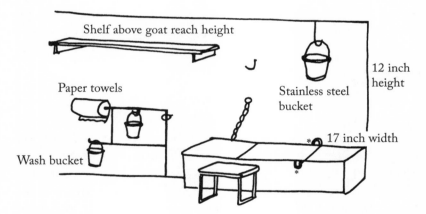

Three slabs 17 inches x 12 inches, on brick, rendered stand, no corners or cracks.
Wash bucket frame set in floor with a ring bolt into the wall, swings out so milker can reach wash, etc. Top bucket for used towels.
A 24 inch chain goes up onto a hook around the doe's neck.

*Rings set in when bail is made for tying back legs when young does first come in if necessary.

Small milking shed.

ing in time — this saves a great deal of trouble. Even on one-goat farms a good milking shed, which can be shut off from the sleeping quarters, with cupboards for minerals etc., makes work much easier.

Milking

Goats are milked differently from cows, both on machines and by hand. If milking by machine, the pulsator must be set at the lowest possible level — milking may take a little longer, but the result is more satisfactory in the long run. If milking by hand, a squeezing action is employed and not a stripping motion as used on cows. If possible always milk a goat before you buy it, some are very tight indeed and therefore good neither for machines or hand milking.

Some does have very short teats when they first come into milk, but do not think that leaving the kids to suckle will help then lengthen — it will not. Hand feed the kids her milk, milking her twice a day and usually, after about three weeks of the full udder forcing the teats to stretch, they will become a decent length. If they do not, seriously rethink breeding from her as neither she nor her progeny will be effective hand or machine milking prospects.

Care of Milk

A dairy inspector told me once that he was taught that germs could swim, but they could not walk. True in essence, if not completely in fact. If a goat's teats are washed prior to milking — the mixture can contain a little mild detergent and disinfectant — then they *must* be dried with a disposable paper towel, a fresh one for each animal. The washing should include forward from the udder along the belly to remove loose hairs, these can get dropped into a bucket or an inflation. Wet, undried teats, whether machine or hand milking, drip a potent brew of coliform and other bacteria either into the bucket or the milk lines. Whether washing and drying is practiced or not, and opinions differ as to its contribution to cleanliness (personally I have always done it), the first squirt of milk *must* be discarded. Tests with dairy cows have shown that an obstinate count of 20,000, chiefly *E. coli*, disappears once this first squirt is removed. The bacteria and dirt in the end of the teat are not removed by washing, so this process is in fact more necessary than washing.

Cooling

Milk must be cooled as fast as possible once it has left the goat, whether half a pint or 10 gallons. If this is not done, all the hygiene and healthy animals will be wasted. In machine milking the milk goes straight into the cooling vat which is usually kept at the freezing point. In hand milking it is run through a ripple cooler as illustrated previously, and then into a container which goes straight into a cool room or vat to be kept at a temperature no higher than 32 degrees. I was warned against using plate coolers years ago by a dairy technician. He said that if the milk was not being collected every day and pasteurized, they are a potent source of bacterial infection. The inner surfaces cannot be properly cleaned as can a ripple cooler's outside surfaces.

Milking Equipment

In hand milking, the milking bucket should be stainless steel with a three-quarter lid to minimize "fall-in." Plastic is not suitable for this operation if the milk is being sold commercially (nor even for the home in my opinion either). Even the highest grade food plastic with a minimum of handling has to be replaced annually or more often as the lactic acid affects the surface. When it was still legal to sell milk from the farm and I had a fairly large turnover, I risked losing customers as I refused to put their milk up in anything other than glass. If buying second-hand milking machines, have the stainless steel checked. Some of the early ones were of poor quality due to economies with cobalt at the time of manufacture. If examined under a microscope they look a bit like a lunar landscape.

Cleaning the Equipment

Everything that is used at milking time must have the first *two* washes in cold water, failure to do this aids the build up of milk stone. The next rinse can be boiling water with or without detergent and the last rinse is boiling water again. A dairy acid cleaner needs to be used once or twice a week. Nowadays organic cattle dairies use peroxide in some cases, which seems to work well.

If the two cold and one boiling rinse are done properly, once a week or less always seems to be enough. In the old days, many dairymen ran boiling water straight through their lines and coolers immediately after milking. Milk stone built up incredibly fast and I bought my six-foot cooler off just such a farm. It took me four hours of hard work to remove the built up milk stone which reached a quarter of the way across the pipes from each side, leaving about half the cooler operative (and dirty) in the middle. Nowadays peroxide and colloidal silver are often used for disinfecting and cleaning and are, of course, completely safe, unlike chlorine and some other dairy cleaners (which are not permitted in an organic set-up).

Health of the Goat

All the measures suggested above are a waste of time if the goats are not healthy. This has been dealt with elsewhere. In Australia, where the calcium/magnesium deficiencies are almost inherent, the addition of minerals to the ration and the suggested feed ratios as set out in Chapter 6 eliminate the risk of mastitis and acetonemia. If there is any doubt about the quality of the milk, have a bowl containing a little detergent and water and squirt some milk into it; it will become viscous if the milk is wrong, even with subclinical infections.

Off Flavors in the Milk

There are various causes for this and it can be a persistent problem in about .05 percent of the goat population. In 28 years of milking I have only met two goats that were incurable; their milk was not palatable, although apparently bacteriologically of good quality. Vitamin B12 injections can sometimes help and an old tip was a teaspoon of sodium bicarbonate in the feed daily, I have not tried it.

It must be remembered that in middle European countries, where goats habitually graze herbs in their feed, a flavor is expected in the milk. I have had more than one of my European friends tell me that our milk is too bland. The taste of goat's milk in their home countries is not, as is often claimed, the result of poor hygiene.

In Australia the main cause of off flavors is a cobalt deficiency, sometimes induced by a sudden change in the quality of the pasture. Giving an intramuscular injection of two ml of Vitamin B12 daily until it clears up usually deals with the problem quite well. The two goats that I could not cure, referred to above (I advised their owners not to breed from them), only responded slightly to vitamin B12. Occasionally too much capeweed will affect the taste of cows' milk but, in spite of grazing it almost exclusively at times (for want of other feed), I have not found that it affected the goat's milk. This could have been because my goats receive the correct supplements.

Pasteurization

This is a process of heat treating milk to kill pathogenic bacteria. It also kills beneficial ones that cause milk to sour naturally with the passage of time and the bacteria that inhibit undesirables. Pasteurization does *not* make dirty milk clean; and in fact the milk that is to be pasteurized should be extra clean because the inhibitory bacteria no longer function. For this reason, if it does become contaminated after treatment, it can become very dangerous for those who use it.

Pasteurization kills the CAE virus which is, and has been, a concern. Home milkers must be sure that their house milkers are free of the disease, as it is milk-transmitted, otherwise the milk should be boiled before use — particularly if it is for an infant.

Correctly pasteurized milk is heated to 165 degrees Fahrenheit (75 celsius), held at that temperature for a minimum of 15 seconds and then immediately cooled in a heat exchanger to 42 degrees F (10 celsius). It is then transferred to an external chiller and held there at 39 degrees F (4 celsius). This is proper commercial pasteurization and the machinery is efficient but expensive; it has to be to lower the milk temperature in a matter of seconds — if the milk is allowed to "stew" it becomes unpalatable and of very low nutritional value.

There is no doubt that pasteurization robs milk of some of its quality. Half of the vitamin C, 25 percent of vitamin A, 16 percent of vitamin B2 and 20 percent of the iodine (by volatilization) is lost in the pasteurization process. With an incorrectly set up pasteurizer the damage would be much greater. But against this we have to set the risk factor of milk-borne disease. Milk that is cleanly gathered at home from healthy goats with minimal handling should be safe. But in Australia, with great distances involved, milk of varying age from commercial goats is handled quite a number of times before the final packaging and placing on the supermarket shelves. It may sit there for a day or two before being transferred to the shopper's car (invariably hot in the summer), while the shopping is finished. Thus it may not reach the home refrigerator for another two or three hours. Raw milk cannot take that sort of treatment and survive.

In the 1980s the results of exhaustive surveys done on milk from a number of outlets in Victoria and New South Wales showed milk with unacceptably high bacterial and *E. coli* counts in many cases. In some samples quite unpleasant and dangerous pathogens were found. As milk is generally pooled at collection, the clean milk also became contaminated and the result would *not* have been safe for human consumption. With further farmer education and stricter controls, it is hoped that the quality of the milk offered to the public will improve. The main result of the survey was to show that pasteurization was very necessary if people's health was a consideration. Hopefully this rule may be relaxed for cheese making and private sales; regrettably the slovenly habits of a few have penalized a whole industry. Absolutely no blame attaches to the doctors (Covenor and Darnton) who did the survey, they had no other option. Chapter 16 gives further information on pasteurization and the qualities of goat's milk.

Products of Goat's Milk

Apart from whole milk sales, goat's milk is made into an almost bewildering array of products ranging from cheeses, yogurt, soap and shampoo to talcum powder. There is a large array of cheeses that can be manufactured from goat's milk and it is also useful for mixing with cow and sheep cheeses. The cheese maker to whom my milk was sold found that by adding it to his cow's milk feta he made a far superior and much sought after product.

For the home cheese maker there are many excellent books on cheese making available today.

Feeding Orphaned Animals

First make sure that the goats used are CAE-free, orphans are already at risk without subjecting their immune systems to a ready-made disease. Undiluted goat's milk seems to be tolerated and assimilated by practically every kind of orphan, from little birds, to kangaroos (apparently cow's milk is *not* suitable for kangaroos as it causes blindness), lambs and foals. Calves do well on it and can occasionally be fed a 50 percent dilution if necessary, the same for pigs — the latter do particularly well if the milk is allowed to sour first. When feeding calves, etc., add a little dolomite to the milk once or twice a week; it ensures healthy insides and lowers the risk of scours (which is basically caused by a magnesium shortfall).

Marketing

In Australia we do not have the close settlement that allows a central marketing system like those I have seen in the United Kingdom and heard of in Europe. Here, anyone thinking about farming goats has to make sure that the farm is set up within reach of a contractor or market. We have no Milk Marketing Board here as of yet — but it is a good idea for start-up operations to have customers close by.

The difference between Australia and France, for instance, was highlighted by an Australian goat breeder who had been working in France. In the Loire Valley alone there were about 40 goat farms. They all produce specialty cheeses and send their products directly to the Paris markets. There are 15,000,000 people in that metropolis and the producers cannot begin to meet the demand. In Australia the enormous distances between centers and the scarcity of the population make marketing a problem.

When setting up a goat farm, check the locality of the nearest contractor and how the pick up (or delivery) system works. It is obviously totally uneconomic to have to deliver the farm's milk 150 miles — when the price is paid delivered — extremely large amounts would be needed to make the operation viable. Many aspiring goat farmers often do not consider this aspect until it is too late.

General

Cooling, cleanliness and the good health of the animals has been covered at length. However there are other factors in the quality of milk, especially if selling to cheese manufacturers. For this trade the milk must not only be top quality — high class cheeses are very temperamental — there also has to be a balance between butterfat and the proteins/solids-non-fat (SNF) levels. Milk with very high butterfat that are not backed up by good protein and SNF cannot be made into good cheese.

A master cheese maker (who started his apprenticeship at age 11, in Italy) to whom I first sold milk for cheese was a good, and strict, teacher. He would wax profane about milk where all the important ratios (butterfat, protein and SNF) were wildly out of balance. He expected (and got) a minimum of 4.5 to 6.5 butterfat with SNF in the range then used of eight to nine. He had me taking the SNF readings correctly so that I could monitor it myself. The bottom line was giving the goats the correct amount of dolomite and minerals in their feed, too little and the SNF went down — anemic goats (lack of copper) and the butterfat fell. He claimed that it was impossible to make good cheese with unbalanced milk. He also believed it was unsuitable for drinking and would upset people's digestion.

For years the emphasis has been on high butterfat, usually with little relation to the amount of milk produced or the SNF and protein levels. Rather a pointless exercise as goat's milk is rarely, if ever, used to make butter.

My type of goat husbandry has been developed to avoid the use of antibiotics in any form, either orally, by injection or intramammary preparations. This is mandatory (in my opinion) when selling milk for cheese. Seeing the cheese maker's despair at having to throw away 200 gallons of prospective cheese because of penicillin contamination was enough to convince anybody who had any doubts on the subject. Cheese starters cannot perform on adulterated milk.

Before I learned how to farm goats without antibiotics and avoid mastitis, I timed very carefully how long after the administration of a blue-dyed intramammary preparation I could see blue specks in the milk. The time was three weeks.

Official Herd Testing and Q Star Tests

This is a rundown of what is done in Australia; most countries have similar systems. In the U.S. this program is called Dairy Herd Improvement (DHI) and involves herd testing of 305 days or less and one day tests.

Herd Testing

This type of official testing is carried out in most countries. The details of management may vary, but the results are much the same. In the dairy cattle industry it has been in use for many years to pinpoint the most productive bulls (and cows). If enough animals are tested, then the performances of the males can be evaluated and compared.

Unfortunately, in the goat industry in Australia there are very few big herds under test, and the numbers so far are insufficient for any worthwhile data. In 1989, the Victorian Herd Improvement Association recorded 50 herds (in four breeds) with an average of five goats per herd. These numbers, of course, are nowhere near enough. Such small herds fall into the hobby or show category, where the management is a far cry from that in a commercial herd which has to make a profit to stay in business.

Hopefully the scheme will eventually be easier to implement, so larger herds will feel encouraged to take part. The in-line meters to measure the milk from machine-milked animals are now available for goats here. Prior to that, those on herd recording had to be hand milked for each test.

The system at present is to test each goat once a month, and multiply the result by 30 to get a mean. Originally the recording was for 365 days, then it was cut back to 280 days to come into line with cow practice. This meant absolutely nothing in terms of telling a prospective buyer or stud breeder whether a goat milked through the winter or not. Fortunately good sense has prevailed and it is now possible to test for the 365-day period once more. The butterfat and proteins in the milk are measured at each test and at the end of the year each goat's full amount of milk (in kilograms, butterfat and proteins) are listed for the owner. At most tests a monitor is necessary if the desired qualifications are to be gained. There is a system operating now where a monitor is not necessary, but the fact is recorded in the results which makes it somewhat invidious — especially if the samples are taken automatically on a milking machine.

A printout is supplied at the end of each goat's lactation giving the animal's production figures. From these results does are given show figures which enable them to take part in Herd Test Type and Production classes. They also supply the basis for the production figures which become part of the goat's history.

Q Star Tests (24-Hour Tests)

These 24-hour tests are held whenever a goat owner books them. In Victoria there is usually a two-month period for booking ahead. Regulations in other states may differ.

For a Q Star test, a sampler and two monitors are required. The goats are milked out the previous night and stripped in the presence of the monitors, who will have checked the identifying tattoos, etc., which is normal practice in any milk test whether 24-hour or herd testing. Exactly 12 hours after the strip-out the previous night, the goats are again milked. The sampler weighs the milk and takes a measured quantity of milk from each doe, puts it in a bottle marked with her tattoo number and leaves the bottle to be filled from the night milking (exactly 24 hours after the first strip-out). The butterfat and proteins are evaluated from this sample by a qualified authority, usually the nearest cheese or butter factory.

The doe receives time points (up to three months) from the kidding date which are taken into account. A doe who is tested more than three months after kidding (sometimes 12 months later on the run-through) may still gain her Q Star although her milk might be a lesser amount than a doe gaining her Q Star early in the lactation. The minimum butterfat allowed is 3.5 and the figures are added up to give a doe her Star, Q Star or Star Q Star qualifications. Eighteen is the minimum number of points required. Figures from good does have been in the high 20s and 30s.

On the basis of the Herd Test and Q Star figures, qualifications are given to does and their fathers (and sons). The system differs slightly in other countries. In the United Kingdom, in accredited shows, no doe may be shown in her inspection class if she has not gained her Q Star the day before the show (the animals have to stay overnight at the show grounds).

Q Stars pass from one generation to the next if the stock is registered (and therefore tattooed). In Australia, the daughter of a Q Star doe that gains a Q Star in her turn is termed Q Star 2, in the UK she is Q Star 1. In Australia a Dam of Merit must have three Q Star daughters, and Sire of Merit five Q Star daughters. A "dagger" buck is a buck whose mother and father's mother both gained a Q Star.

Section marks, numbers and figures are the qualifications for Herd Tested does and their sons. These are always put before a goat's name, as is the dagger for bucks. Q Star qualifications are always placed after the animal's name.

Health Benefits of Goat's Milk

Dr. Bernard Jensen is a clinical nutritionist who with Mark Anderson wrote a chilling book on modern foodstuffs called *Empty Harvest*. In this book he points out that each generation is slightly less bright and less strong than the preceding one due to our current food supply which is grown on poor land and denatured by processing. He argues that we are the products of our food supply. He has also written a definitive book on goat's milk — possibly the most up-to-date book we have yet had. *Goat Milk Magic: One of Life's Great Healing Foods*, is the name of his self-published book on the subject. On page 334 is a table taken from the book which describes the nutrients found in goat's milk compared to breast and cow milk and includes vitamins as well as minerals (all nutrients are quantified per liter/quart).

Dr. Jensen emphasizes what has been said many times before, that the goat is a very particular feeder who will go to great pains to get the minerals and nutrients it needs, more so I think than any other animal. Consequently, the milk of the goat is high in the minerals and nutrients it consumes.

Composition of Milk (per liter/quart)

Item	Human Milk	Goat's Milk	Cow's Milk
Energy/E calc.	710	620	660
Protein/gm	11	32	42
Fat/gm	38	40	37
Carbohydrates/gm	68	46	49
Calcium/gm	340	1,290	1,430
Magnesium/mg	40	10 to 145	120
Phosphorus/mg	140	1,060	1,120
Sodium/mcg	7	15	27
Potassium/mcg	13	46	45
Iron/mg	3	1	0.5
Zinc/mg	3.5	2.4	3.5
Chloride/mg	75 to 450	1,200	1,050
Vitamins			
A/IU	2,000	2,074	1,500
B1/mg	0.160	0.400	0.440
B2/mg	0.360	0.1840	0.2100
B3/mg	1.47	1.9	1.0
B5/mg	1.84	3.4	3.5
B6/mg	0.100	0.70	0.640
B12/mcg	0.03	0.6	4.3
Folacin/mcg	52	6	55
Biotin/mcg	8	39	31
Choline/mg	90	150	121
Inositol/mg	330	210	110
C/mg	43	15	21
D/IU	22	24	14*
E/IU	1.8	—	0.4
Essential Fatty Acids			
(per 100 gms of milk fat)	4.1		2.6
Fat globules of less than 3 dram (percent)		63	43

* not fortified.

Note: Work done by Drs. Archie Kalokerinos and Glen Dettman established that in human mothers deficient in vitamin C, the content of that vitamin in their milk was virtually nil.

Dr. Jensen devotes a chapter to the feeding of babies and their requirements. Another chapter covers the kind of sodium found in goat's milk and explains why it is so good for us and our children — and it has absolutely nothing to do with table salt. His ideas on goat feed are remarkably similar to mine and he places great importance on magnesium and sulfur.

Dr. Jensen calls the goat the "poor man's medicine chest" and heads the chapter with the following,

> 'I do not know how many times I've driven down country roads, spotted a goat in a pen or barnyard, and stopped to see why the people had it. I heard some wonderful testimonies."

Further down the page he says that for those recovering from devastating disease there is nothing like warm, foaming milk straight from the goat to restore strength (remember Johanna Spyri's *Heidi*). The vital energy of the goat remains for about three hours. Then it is gone. I can only suggest that those who are interested in goats and goat care should send for Jensen's book.

Pasteurization and Health

I have already mentioned that in Australia unpasteurized milk cannot be sold. About 20 years ago Dr. D. Weston Allen, M.D., wrote an article for the *Australian Goat World* called, "In the Interests of Better Health." It pointed out the disadvantages of pasteurization and seven names were listed in his references.

In answer to the question: Does pasteurization affect the nutritional quality of milk? He wrote,

> "Fresh cow's milk contains 20 to 25 mg per liter of vitamin C — half of this is destroyed by pasteurization. It also destroys 25 percent of vitamin B1, 9 to 16 percent of vitamin B2, and 20 percent of the iodine (by volatilization). The composition of the calcium in the milk is also changed so that six percent less is available."

He stated that laboratory rats fed exclusively on pasteurized milk for several generations show degenerative changes and lose the ability to reproduce. This does not happen to animals fed raw milk.

I know many people these days do not want to be tied down to milking a goat. I am profoundly thankful that I had goats at the time when my children were growing and my husband was sick. Like all young, my children suffered a variety of ailments, some more serious than others. When my eldest daughter (aged 4 years) was horrendously scalded after trying to lift a boiling kettle *over* the fire guard to help with the washing up, the doctor told me the scars would be life-long. She, like the other four children were on ad lib goat's milk. Five days later the doctor called and said I could get her the next day. He was totally mystified as to why she had healed so quickly without any scarring and did concede that maybe the goat's milk had something to do with it.

There are innumerable anecdotal cures attributed to goat's milk, and I have no doubt that they really happened. I saw some quite incredible turn arounds in children's health. Perhaps the most happy one was a child whose mental levels were well below what they should have been at age three. She had been unable to assimilate the required nutrients from cow's milk to attain the intellectual level for her age. After six months on goat's milk, she caught up to her age group and she never looked back.

But there is always the chance that children or adults who are allergic to cow's milk will be the same with goat's milk — it can and does happen; I always warned parents with sick children to wait and see.

Still, the bottom line in this country at present is that pasteurization of commercial milk is mandatory, but the above should encourage those with the facilities to get themselves a goat to make sure that their children at least receive the best milk available. It is critical to make sure that the goat is from CAE-free stock and in good health.

Chapter 14
Goats for Meat and Skins

Breeding Meat Goats

Boers

The advent of the Boer goats in Australia has given the meat industry the incentive needed to get into gear — at last. The cost of their introduction made them expensive initially; but now, 10 years later, they are very much part of the goat keeping scene. Their docility makes handling and farming them much easier than fiber and dairy goats which has been an encouraging factor. Broadacre farmers are now considering them as an adjunct to their present enterprises. There are two normal colors in the breed, the black/brown and white, where the color is about the ears and head and the red Boers which are the same color all over.

A massive upgrading and embryo transplant program has now increased the number to viable proportions following their original arrival at Terraweena around 1990. Some came directly from New Zealand. Upgrading has already been covered in Chapter 5. The Boers are being used widely in crossbreeding operations and should start returns to farmers fairly soon as a source of meat and not just for building up stocks.

Boer goats (Angora in background).

Ferals

Prior to the arrival of the Boers, the meat market was supplied by feral goats and few people made any intelligent use of even that market. A friend in Western Australia who had some land around Meekatharra went up and shot out as many of the feral bucks as possible and then introduced low appendix, medium-sized Nubian bucks from sound, tough lines. The progeny were ready for market at three months and sold very well. Using bucks that are too large can lead to both mating and kidding problems. Boer goats or crosses would have been even better had they been available. However there has been, up until the advent of the Boer, no real program to supply a very keen meat market.

A few abortive efforts to bring in semitrailer loads of ferals were not well thought out. Of two attempts I knew of, one was to deliver them to the abattoirs at Sunbury, Victoria (set up for large animals), and the entire load roamed the district for next few months because cattle yards do not keep in feral goats. The other was a would-be meat "king" who came unstuck because of refusing to believe that goats needed feed and minerals. He was sure that it was not necessary to give them anything except the grass in the starved (and probably minerally deficient) paddock into which he drafted them, in spite of being warned by the vet and myself. He lost 50 percent of the animals in a few days.

The feral herd in Australia has been an untapped genetic bank for most breeds of goats. Some obviously carry the genes of Angora, Cashmere or milk, but others are patently fast-developing meat producers and should cross well with the Boers. They need to be carefully selected according to horn type and body frame, etc.

One entrepreneur who operated for a while was shipping animals over to the Middle East in converted jumbo jets. They were only 14 hours in the air — often less than the time they had spent in transports coming to the holding paddocks. For the local meat trade, except in rare instances, fiber and feral goats are preferred because they are easier to skin and the source is more reliable than depending on dairy culls.

Meat Industry

The meat industry is one area where hybrid vigor can be used to its greatest advantage. For a program of that kind, the same does can be used for as long as possible and mated to totally unrelated bucks. Observation and careful selection of the base feral does will pinpoint those whose stock mature the fastest and a good herd can be built up.

There is a good demand for goat meat, both on the hoof and dressed, worldwide, particularly in Arabian and Asian countries. At a seminar held in Queensland, John O'Gorman of the Australian Meat and Livestock Corporation gave an excellent presentation on the need for a steady supply of goat meat, both in Australia and among our northern neighbors. When discussing the supply with him afterwards, he was so discouraged by the total lack of serious response in this country that he was thinking of giving up trying to organize anything.

Ferals that are brought in for the purpose of breeding and fattening must be handled very carefully. They have roamed at will over vast areas and have never been short of the nutrients they need. It is imperative that they are given good mineral licks and supplementary hay if necessary, according to the time of year. Concentrate feeding would make the venture uneconomic, unless seconds from the harvest could be obtained cheaply enough for them to be a viable option.

The most economic type of lick would be that described in Chapter 6. In the short or long term, that mixture should cover everything needed for lactating does and fattening youngsters — provided they have hay.

Kid Meat Market (Capretto)

Dairy kids *do* make very good eating. Usually in Australia their disposal is a private matter, or the owners put them in the freezer for the year's meat. There is a small gourmet trade for well-reared dairy kids and the farms usually have arrangements that go on from year to year — if their product is good enough. These kids are not very popular with large-scale meat exporters and processors here due to the uncertainty of the supply.

In Europe, as here, there is a gourmet trade in well-grown milk-fed dairy kids. In the Loire Valley, mentioned in the last chapter, there are a great many goat farms in a concentrated area so kid rearing enterprises are profitable. One such farmer at Lusignan reckoned to handle 42,000 kids in four months each year, they are fed goat's milk from a calfateria. The goats supplying the milk are poor milkers who only keep going for four or five months, but supply plenty for the kids in that time. The kid-rearer picks up the kids and never has to travel more than 25 or 30 miles to do so. One unit that I have seen in Australia ran about 40 kids to a shed, on a four to eight teat calfateria (with ramps round it for the kids to reach the teats). It was cleaned out daily and refilled when necessary, milk replacer was used and once the kids had been taught where to find the milk, they did well. There was no pushing or bullying, probably because the milk was always available. Soured fresh milk also works well for this sort of operation — it is easier to digest.

The difficulty in Australia is again distance. Every so often there are reports of new abattoirs opening for the kid meat or export market. Again fiber goats or ferals are preferred because they are easier to skin, also they are not so "cute" as dairy kids, which sometimes upset the abattoir workers. The introduction of the Boer goats and a regular supply of meat animals will mean that the markets can stabilize and eventually handle dairy kids as well.

For as many years as I have been in goat farming (35 at the time of this writing), the wastage in unwanted kids has been one of the grey areas.

Prices for meat goats are not very high, but the costs of rearing them for the required six or more weeks can amount to $40 worth of milk against the $20 offered per kid. It is obvious that it is cheaper for a commercial milking farm to bury the kids at five days. If there was a market for kids at that age it would be worthwhile. I found my own markets and was never reduced to using the compost heap. The kids were bought at five days by French people and the $10 I was paid per kid was quite enough. They bought enough from me to keep their freezers supplied for the year.

Killing for Home Consumption

With goat meat it is quite difficult to determine if the animal has been well fed and reared. Many years ago a butcher who regularly bought my culls for pet meat, took a goatling that had sustained an injury off my hands. She was a great pet, so we could not eat her ourselves. The butcher's family ate the doe and to this day labor under the impression that it was extra good two tooth lamb. Unwanted males do not need to be castrated if they are to be killed by the time they are three months old, in fact, some gourmet outlets prefer them entire.

Shooting Goats

It is necessary at times to shoot goats, either because they are suffering or for home meat. It is less traumatic to tie them to a favorite source of food, approach quietly from the rear and shoot them where the spine goes into the head. This works equally well whether the goat is polled or horned. Shooting a horned goat from the front can sometimes be difficult because of the thick horn area. The gun should be held at the same angle as the head.

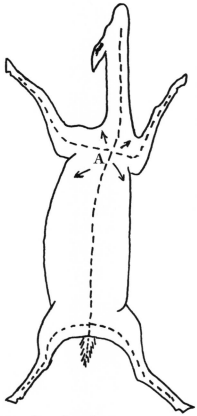

A= the starting point for a clean skin

Where to start skinning.

Skinning and Butchering Goats

Goats, even more than most animals, are easier to skin while still warm — they are *very* hard work when cold. Tie the goat's back legs to the ropes and pulleys and raise the carcass to a convenient working height. To skin, commence along the dotted lines, for a clean skin that is free of flesh, start at the brisket "A," and work toward the rear end and backbone, punching the skin from the ribs and stomach. Cut around the anus and along the underside of the tail; to skin the tail, grasp the skin in the hip area and pull. Then proceed to pull along the neck towards the ears, if you start at the other end the whole skin will be covered with a sheet of flesh (muscle), which makes curing the skin *very* hard work.

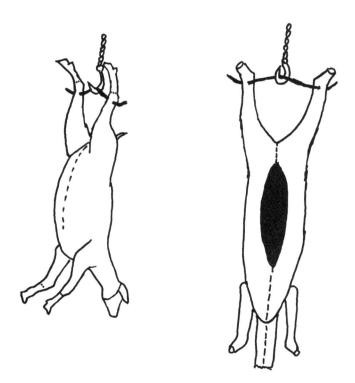

Butchering and skinning.

To clean the carcass, cut off the feet below the knees and hocks. Split the stomach from the end of the breastbone to the pelvis and extract the intestines. Leaving the carcass to set, if a cool fly-proof area is available, also makes the next operation easier. I did not have such a place and Owen Dawson, who provided much of this information, suggests that in that case go straight on and cut up the carcass. Split the crutch and the full length of the breastbone with a cleaver. Cut around the anus with a knife to extract the bowel and bladder. Split the neck with a knife, grasp the lungs and draw downwards getting them out complete with the gullet.

The Skin

Lay the skin out flat, hair downwards and liberally cover it all over with salt, fold the sides in, so it is salt to salt then fold together and hang in a porous bag (woven plastic or jute bags work well). Leave it for three to four weeks before starting the next stage, but it can be left longer. To cure skins, purchase John Leidreiter's skinning book, *A Handbook on Knives, Skinning & Tanning* (if you can find it) and outfit, directions are very easy to follow and the finished result is most satisfactory. An electric sander to "finish" the skin is also very helpful.

In countries where goat meat is a normal part of the diet, the skins are used for making high-grade leather gloves, clothes and artifacts. Australia imports beautifully treated goat leather for the *haute couture* market from Italy — perhaps one day we shall produce our own.

All ages of goat and kid skins make beautiful rugs, throws or clothes. I have the skins of most of my favorite goats that had to be killed in the CAE era as chair seats and rugs, at least it is something to remember them by. My most unusual request was for a goatling skin to make bagpipes; it had to be skinned *in toto*, but apparently it was most successful, although I never saw the finished product.

Chapter 15
Showing Goats

Taking a goat (or goats) to shows can be a most rewarding and entertaining occupation for all the family — young and old. It is also a good way of teaching the younger members of the human family to behave well whether they win or lose.

Showing should not be taken too seriously. If the children actually help with or are part owners of the goats, they feel involved. This worked well with mine; they had to give a hand in the goat house when convenient, collected their goat's prize money and if it was sold, half the proceeds went to their bank accounts. Agricultural shows have many sections and each member of the family can find something to exhibit. We would take off with the goats, generally does, handicrafts, one son's prize rooster and wives (washed and dried in front of the slow combustion stove the previous evening), even sometimes the other son's stationary engine and occasionally bucks too. It provided great opportunities for family outings.

Regulations

Different countries have different regulations governing goat eligibility for shows. In Australia they must be registered with the Dairy Goat Society of Australia, the owners do not have to be members, although they generally end up joining. This, of course, includes IR (Identification Register) animals and these may compete against other similar goats; according to the constitution they are eligible to compete against herd book and the appendices for the milk classes, i.e., best udder, Q Star and herd recording. IR Bucks can only compete against others of the same if there are classes for them. Not all shows in Australia hold classes for IR goats, although their equivalent are shown in the United Kingdom.

When buying the family milker, make sure you get the papers which consist of a registration form and transfer — now usually amalgamated into one. There must also be proof that the goat is free of CAE, and occasionally Johne's disease. Those who have accredited herds for both can show according to the regulations. Goats from small herds may have to be individually tested, this is fairly expensive in some states.

The first step is to obtain a schedule. The goat classes are usually fairly clear. There will generally be a closing date for entries, see the forms are sent in on time. This makes the convener's task much easier — although some shows accept entries on the day of showing, but not all. Enter the goat in its appropriate classes and read the regulations carefully. Usually one is asked to wear white clothing or a dust coat, which helps to show the goat off to the general public.

Attitude to Showing

A word of warning about shows, if it becomes a matter of life and death to win and your whole day is ruined without that coveted first place ribbon — forget it. Judges vary and every one of them can look for slightly different points. Personally, having had to earn my living with my goats, I rate milking ability as being very important. No matter how beautiful she may be, if a goat falls down in that respect I do not consider she is worthy of high honors.

But generally a study of show results, under widely differing judges, shows the same names occurring again and again in the first three or four places (the order changes occasionally). Now and then a new one comes on the scene, which I always find very pleasing.

It *is* nice to take home a few good prizes and to compare one's goat with the others at the show. If you feel the judge has made a mistake and wish to query it, do it after the show, quietly, without causing a disturbance. Judges are human and I have been at the receiving end of a couple of real bloomers. I leave it at that, after all, I know the quality of the animals I show.

Show Requirements

> White coat or clothing.
> Schedule, with your classes marked (keep a copy of the one you sent in, and have the birth dates of your goats on it). Some shows have catalogs but not all.
> Chain collars (as for dogs) — they look better.
> Leads — you can plait fancy ones — with spring clips.
> A body brush.
> Scissors.
> Light smart dust sheets to wear between classes, dark-colored goats can show the dust (optional).
> Short (about 8 inch) lengths of chain with a swivelled spring clip on each end — one for each goat. Tying your goats up with rope is all very well, but you can guarantee they will have tied inextricable knots in it when you are in a hurry for the next class.
> One long length of chain which can be fastened to stakes driven into the ground in case there are no rails or pens. Not usual nowadays but the norm in the past.
> Milking bucket, washing water and paper towels (will be needed for the children too).

Portable milking bail if you have one.
Hay.
*Feed, chaff and bran — no grain.
Branches, in bundles.
Feed and water buckets, sometimes helps if they have a spring clip on the handle.
Chairs.
Medicine chest. Vitamin C and B12 injections, dolomite, bandages, Arnica, magnesium orotate, etc.
The family's food and drink — very important, you gain no marks forgetting it.

*It is wiser to feed easily digested food at shows. The travel and excitement stress the goats slightly and they do not need to have to digest grain as well.

Preparation

Two or three days before, see that the feet are clipped right up to date. *Never* leave this until the last moment, one wrong cut and you have a lame (and unshowable) goat. Trim the udder (before milking), the hairs on the inside of the ears — only necessary on Swiss breeds, Nubians have hairless ears — round the tail, and along the belly. Do *not* shave the belly; the goat could get chilled and it does not

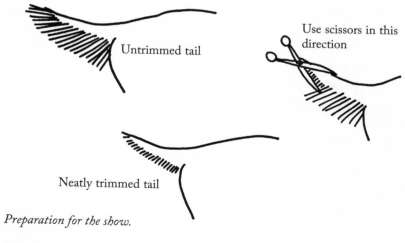

Preparation for the show.

look very attractive. If you are not experienced with clippers leave well enough alone — nothing looks worse than a goat with clipper bites out of it everywhere. Personally I do not like the look of a clipped goat, it is much done in the United States, but a full coat in top health takes a lot of beating. From the judge's point of view it also feels better.

The Day Before the Show

Have a good horse or dog shampoo ready or failing that, a tablespoon of white Handy Andy (cloudy ammonia), the same of methylated spirits and the same of eucalyptus in a bucket of hot water. This makes an excellent shampoo (I've won countless shows with it).

Before washing have a rug ready for each goat (like the one in the drawing *Aids to showing* — next page), this can be made from an old towel. A flap can be left down one side which is passed under the belly and sewn up on the other side. Restrain the goat in a bail if possible, the all-purpose one illustrated earlier is ideal. Wash the goat thoroughly with warm water using a plastic brush to remove the dirt on the sides, pay particular attention to under the tail, around the scrotum in bucks, the knees and the feet. Rinse it off well — the whole operation should *not* have taken longer than five minutes — the animal is not supposed to get pneumonia. Have the wrap-around rug ready to put on immediately, so they do not have a chance to shake and disturb their nice smooth coat. This under rug can be covered by a smart one and the goat is then turned out to graze as usual. These rugs should not be removed until the goat is about to go into its class — cut the stitching, slide the rug off over the goat's tail and lead a gleaming, clean animal into the ring. The metamorphosis can be quite astonishing.

Bucks of course cannot wear a rug after washing that goes under the belly. They will need an apron as shown in the illustration. Tie it firmly round the middle, in *front* of the penis, so it cannot spray its front legs and make them all sticky and dirty again.

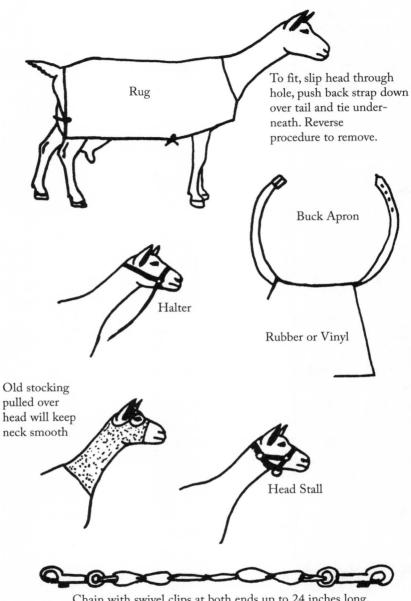

Aids to showing.

Many people have complained to me that my show preparation is too difficult. I can only say that when I was showing seriously (and winning — I had to make it pay or we could not do it), I had a large young family, a not too well husband and *no* electricity or hot running water — it can be done.

If the show is more than an hour from home or the weather is very hot, it is a good idea to give each show animal a teaspoon of vitamin C and the same of dolomite powder the previous evening. Add a crushed up vitamin B6 tablet as well if the goat has ever shown signs of car sickness — this will ensure that they do not arrive at the show in a state of collapse. If it is a very hot day and they do not look happy, crush up two 400 mg magnesium orotate tablets and just pour them into its mouth — instant recovery — this mineral is absorbed through the mucous membranes. There is no doubt that temperatures over 100 degrees deplete magnesium in any sort of body. Harold Willis, when I discussed this with him by letter, confirmed that it did much the same to the soils — too hot for microbes who go down deeper and take the goodies with them. A fact I had long suspected.

The Day of the Show

These guides may be different in the United States, but the following apply in Australia and serve as an example of show preparation anywhere.

Do not milk your goat on the morning of the show, they have to be shown with full udders. No one told me this the first time I showed my goats. Nor should the travellers be given concentrates before the journey. Hay to eat in the van is alright.

The goats will still be nicely sewn up in their rugs, hopefully the kids too (but they are great rug removers), so no brushing will be necessary, perhaps a quick wipe for the feet will be all that is needed. Once at the show, drive your pegs, attach the goats and give them a hay net and some water and branches. No hard feed until after they are milked out. Collect your catalog and numbers if any, and tell the steward if any of the goats have been scratched.

Yours truly with an Alpine doeling at local show.

In the Ring

Lead your smart looking goat with the right hand as close to the collar as possible, unless it is very well trained. Remember, from the moment you enter the ring you are on show; have your goat standing well at all times, even if the

Clean, rugged and ready to show.

parade has not started. The steward will tell you the order and when to start walking around the show ring — generally clockwise. When you are called in, stand the goat up as straight and still as possible. Do *not* hold its head too high or its back (however good) will tend to dip.

The judge will move around wanting to see the goats from all angles, so make sure your goat is always between you and the judge. Do not become involved in long conversations with your fellow competitors (even if you haven't seen them for ages) or you will not hear what is going on, muttered asides are the order of the day. Should the judge ask you to walk your goat away and back again, make sure that you keep the goat between you and the judge when you turn.

Cashmere goats and their owners anxiously await their turn in the ring.

When you are called in, try to preserve a neutral expression regardless of where you are in the line up. Thank the judge if you gain a ribbon and congratulate the winner (if it isn't you).

If your doe is in the best udder class you will have to milk her out. Remember the public are watching and they have some funny ideas about goat people. So make a point of washing and drying the udder first and do it properly.

Do *not* feed the milk back to the doe; it's not a good practice at the best of times, but certainly gives a bad impression if done at a show. Feeding milk back eventually ruins the doe's digestion — as adults they are grazing animals not milk drinkers. Curiously enough two does I know of that acquired Johne's disease were heavily supplemented with goat's milk as adults; it must have upset the mineral balance in the gut.

Try to be ready for your classes on time. This can be difficult sometimes if you are showing several animals, but this is also where the children can be a great help. It is difficult to run a show if the competitors cannot be persuaded into the ring on time.

Most agricultural shows have a grand parade which is their way of showing the farm animals to the general public. *Always* go into the parade — it is part of the show and people like to see goats walking round the arena with their ribbons on.

Winning line up at a country show.

Hobby Goats

Many people with hobby goats include showing as part of the deal. There are others who just like goats and having a supply of good, fresh milk for their own use. The goat becomes the family pet, superseding a dog in some cases.

One family I knew regularly took their goat on outings and holidays. The order of the day was the male head of the family, next came Cocoa, who was a very fine numbered Toggenburg only one generation from the original imports, then the female half of the family, followed by the children. If anyone tried to change the order Cocoa didn't like it; she also took no nonsense from the young. Apparently the smart ladies in Hyde Park, London no longer walk poodles, they have miniature goats instead.

For goats in this category, the more tractable and well reared they are the better. It is probably better to obtain the goat as a kid and hand rear her, but remember she is going to be part of the family so do not let her learn bad habits — like getting into the garden.

Looking after a goat is great training (unconscious) for the young in handling animals — they learn about birth and death as a matter of course. All my children at various times helped deliver kids if I needed another hand and would on occasion manage to do so on their own (telling me all about it afterwards with great pride). My neighbors did *not* always regard this as part of proper training for the young.

The children should be encouraged to help around the goat house when young. This way they realize that this is a chore that has to be done, *every* day — animals cannot be left untended. It also makes them realize that if they want a pet, they must help look after it. This works both ways, *you* wanted the goat and you cannot expect them to look after it entirely, otherwise you will turn them against goats, and possibly other animals, for life.

Driving Goats

In the 19th century, goat (and dog) carts were a common sight in most country districts, worldwide. They were used for gathering wood, delivering supplies or goat's milk (on the hoof). They were generally managed by children. On the gold diggings in Australia there were even goat cart races (hence the Billy cart). Unfortunately abuses occurred —

children being allowed to inflict pain and distress on driven goats brought about a prohibition. Eventually in Australia both kinds of traction power (dog and goat) were outlawed.

However there are a smart few goat outfits here now and as long as the goats are properly treated there should be no problems. But officially, harnessing dogs or goats was illegal here, so perhaps special permission must be obtained first. The public like to see them and nothing looks nicer than a well turned-out goat in a smart rig.

Toggenburg goat and cart — the children love it.

In the goat sections of competition in the United Kingdom there are classes for harness goats, turnout and performance, at most agricultural shows. They are a great favorite with the public and, once they have been judged, the owners often give rides to the children. Some carts are obviously homemade and others just as evidently the meticulous work of coach builders. The harnesses are the same — either professional work or homemade — breastplate harnesses are the norm. Some bridles have bits and some are without, but generally the goats are driven from a head collar bridle or led.

The carts I saw there were either two wheeled — usually bicycle type — and similar to a trotting sulky, or they had four wheels (pram wheels were used here). Most were unsprung and the bodies made from particle board or marine ply on a light metal frame. Some were expensive copies of horse-drawn vehicles.

Single wethers were mostly used in harness and I saw one tandem (one goat behind another), another cart had the family milker in the shafts and another had a pair — one each side of the central pole. There are a few people here who bring very well turned out goats in harness to the shows now, too, which is good to see.

In Australia, stories from the gold fields suggest that the local billy was often the cart puller. Goat carts could become as big a draw at shows as they are in the United Kingdom and would not need so much room to transport as horse buggies.

A fancy, billy-drawn cart.

Bibliography

Acres U.S.A. Various issues. Available from: P.O. Box 301209, Austin, TX 78703-0021, 1-800-355-5313.

Adams, Ruth. *Complete Guide to all the Vitamins*. New York: Larchmont Books, 1971.

Albrecht, William A. *The Albrecht Papers*, Volumes 1-7. Austin, TX: Acres U.S.A., 2005-2012.

Allenstein, L.C. "Cowside Practice, molds are causing feeding problems." *Hoard's Dairyman*, 1993.

Auerbach, Charlotte. *Science of Genetics*. Revised edition. London: Hutchinson Press, 1969.

Balfour, Lady E. B. *The Living Soil and the Haughley Experiment*. Revised edition. New York: Universe Books, 1976.

Begley, Sharon. "The End of Antibiotics." *Newsweek*, 1994.

Bellfield, Wendell O. "Megascorbic Prophylaxis and Megasacrobic Therapy. A New Modality in Veterinary Medicine." *Journal of the International Academy of Preventative Medicine*. Volume 11, No. 3.

Biffen, Ray. Talk on feeds given at Pakenham, Southern Victoria, Austraila, 1986.

Blood, D. C. and J.A. Henderson. *Veterinary Medicine*. London and Sydney: Baillierre Tindal, 1979.

Boade, Alma M. and Helen Brooks. *Angora Goat Husbandry*.

Privately Published by the authors, 1972.

British Goat Journal. Various issues are useful.

Caldwell, Gladys. "Work on Fluoride." Reported in *Acres U.S.A.*, March, 1989.

Cawthorn, R.J. and D. Naff. *Observations on the Epidemiology and Control of Parasitic Gastroenteritis in Goats.* Publication of the Central Veterinary Laboratory, New Haw, Weybridge, Surrey, United Kingdom.

Commonwealth Scientific Industry Research Organization (CSIRO). *Rural Research Bulletins.* Various issues.

Davis, Adelle. *Let's Get Well.* 6th Edition. London: Allen and Unwin, 1972.

de Bairacli Levy, Juliette. *Herbal Handbook for Farm and Stable.* London: Faber and Faber, 1952.

Devandra, C. and M. Burns. *Goat Production in the Tropics.* Publication of the Commonwealth Agriculture Bureau, 1970.

Everist, Selwyn P. *Poisonous Plants in Australia.* Angus and Robertson, 1981.

Farmers Weekly, Anthelmintic Spupplement. London: May, 1997.

Fukuoka, Masanobu. *The One Straw Revolution.* Emmaus, PA: Rodale Press, 1978.

Fukuoka, Masanobu. *The Road Back to Nature: Regaining Paradise Lost.* Japan Publications, 1987.

Gerson, Max, M.D. *A Cancer Therapy. Results of Fifty Cases.* 2nd Edition. Delmar, California: Totality Books, 1958.

Glass, Justine. *Earth Heals Everything: The Story of Biochemistry.* London: Peter Owen, 1958.

Goodman, Louis S. and Alfred Gillman. *The Pharmacological Basis of Therapeutics.* 5th Edition. New York: Macmillan Publishing Inc., 1970.

Gregg, Charles T. *The Plague, an Ancient Disease in the Twentieth Century.* Revised Edition. Albuquerque: University of New Mexico Press, 1986.

Grieve, Mrs. M. *A Modern Herbal.* London: Penguin Books, 1976.

Guss, Samuel E. *Management and Diseases of the Diary Goat.*

Scottsdale, AZ: Dairy Goat Journal Publishing Corp., 1977.

Hill, Dr. David. *A Literature Review: Colloidal Silver Uses.* Rainier, Washington: Clear Lakes Press, 1998.

Howard, Sir Albert. *Agricultural Testament.* London: Oxford University Press, 1943.

Hubberd, C. E. *Grasses.* London: Penguin Books, 1972.

Hungerford, T. *Hungerford's Disease of Livestock.* 8th Edition. Sydney: McGraw-Hill Book Company, 1951.

Hunter, Captain John. *Transactions at Port Jackson and Norfolk Island.* Piccadilly, London: John Stockdale, 1793.

Jarvis, D.C. *Folk Medicine.* London: White Lion Publishers, 1958.

Jensen, Bernard. *Goat's Milk Magic, One of Life's Greatest Healing Foods.* Escondida, CA: Bernard Jensen, 1255 Linda Vista Drive, San Marcos, CA 92078.

Jensen, Bernard and Mark Anderson. *Empty Harvest.* Garden City, NY: Pavely Publishing Group Inc., 1973.

Johnson, Clarence. "20 Day Shelf Life in Fluid Milk." *American Dairy Review*, July, 1979.

Kalokerinos, Archie. *Every Second Child.* Sydney: Thomas Nelson, Ltd., 1974.

Kalokerinos, Archie, G. Dettman and Ian Dettman. *Vitamin C, Nature's Miraculous Healing Missile.* Queensland: Veritas Press, 1993.

Kessler, J. "Elements mineraux chez le chevre, donne de base et apports recommands." Paper presented at ITOVIC INRA International Symposium on Feeding Systems for Goats, Tours, France, 1981.

Kettle, P.R., A. Vlassof, T.C. Reid and C. T. Horton. "A survey of nematode control measures used by milking goat farmers and of anthelmintic resistance on their farms," *New Zealand Veterinary Journal*, 1981.

Kinsey, Neal and Charles Walters. *Hands-On Agronomy.* Austin, TX: Acres U.S.A., 2009.

Klenner, Frederick R. "Observations on dose and administration of ascorbic acid when employed beyond the range of a vitamin in human pathology." *Journal of Applied Nutrition*,

Winter, 1971, pages 61-87.

Lamand, M. "Metabolisme et besoins en oligo-elements des chevres." Paper presented at ITOVIC INRA International Symposium on Feeding Systems for Goats. Tours, France, 1981.

Lloyd, Sheelagh. *Control of Parasites in Goats.* Department of Clinical Veterinary Medicine, University of Cambridge, 1980.

Marsten, Hedley R., and B. Robertson. "Utilisation of Sulfur by Animals." CSIRO Bulletin, No. 39, 1928.

McCabe, Edward. *Oxygen Therapies, A New Way to Approach Disease.* Morrisville, NY: Energy Publications, 1988.

McKenzie, David. *Goat Husbandry.* London: Faber and Faber, 1957.

McKenzie, Ross A. and Ralph Dowling. *Poisonous Plants, A Field Guide.* Information Series No. Q192035. Department of Primary Industries, Universty of Queensland, Brisbane, Australia, 1993.

MacLeod M. *A Veterinary Materia Medica and Clinical Repertory.* Saffron Waldon, UK: C. W. Daniel and Company, Ltd., 1995.

Moore, James. *Outlines of Veterinary Homeopathy: Comprising Horse, Cow, Dog, Sheep and Hog Diseases, and Their Homeopathic Treatment.* 7th Edition. London: Henry Turner and Co., 1874.

Moskowveitch, Richard. "Immunizations, A Dissenting View." In: *Dissent in Medicine, Nine Doctors Speak Out.* Chicago: Contemporary Books Inc., 1984.

Neilsen, Forrest H. "Boron, an Overlooked Element of Potential Nutritional Importance." *Nutrition Today,* publication of USDA Agricultural Research Service, Grand Forks, North Dakota, January/February 1988.

Newman Turner, F. *Fertility Farming.* Acres U.S.A. Austin, TX: 2009.

O'Brien, Anita. "Withdrawal Time, What Is It?" *Australian Goat World*, April, 1992.

Owens, Nancy Lee. "Illustrated Standard of the Dairy Goat." *Dairy Goat Journal U.S.A.,* 1986.

Passwaters, Richard A. *Selenium as Food and Medicine: What you need to know*. New Canaan, CT: Keats Publishing, 1980.

Pegler, Holmes S. *Book of the Goat*. 9th Edition. London: Bazaar, Exchange and Mart Ltd., 1926.

Reynolds, E. F. *The Extra Pharmacopoeia*. 28th Edition. London: The Pharmaceutical Press, 1982.

Russel, Mark, Donald Scott and William Hope. "Moldy Corn Poisoning in Horses." Reported in *Acres U.S.A.*, Austin, TX: February, 1995.

Schuessler, W.H. *Biochemic Handbook*. Thorsons, 1982.

Shepparton Veterinary Clinic Newsletter. Information on anti-inflammatories. Victoria, Australia. April, 1997.

Silver Education Coalition. *Colloidal Silver Handbook*. Salt Lake City, UT, 1971.

Stone, Irwin T. *The Healing Factor, Vitamin C Against Disease*. New York: Grosset and Dunlap, Publishers, 1972.

"Technical Data Sheet: Report on H_2O_2 and Colloidal Silver." *Sanitation of Stock Drinking Water*. Australia: Integrity Products, Ltd.

Thompson, George and M.S. Fayette. *Angora Goat Raising and Milch Goats*. Chicago: American Breeder Company Press, 1903.

Townsend Letter for Doctors and Patients. U.S.A. 1989.

Udzal, F.A. and W. R. Kelly. "Entero Toxaemia (Pulpy Kidney Disease) of Goats." Paper presented at *Seminar '96*, Queensland University, Banyo Conference Center, Brisbane, Queesland, 1996.

Voisin, André. *Grass Productivity*. Washington, D.C.: Island Press, Washington DC., 1959.

Voisin, André. *Soil, Grass and Cancer*. Austin, TX: Acres U.S.A., 2000.

Volker and Steinberg. "The vitamin requirements of goats." Paper presented at IROVEC INRA International Symposium on Feeding Systems for Goats. Tours, France, 1981.

Wallach, Joel and M. Lan. *Let's Play Doctor!* Double Happiness Publishing Company. 1997.

Walters, Charles. *Eco-Farm — An Acres U.S.A. Primer*. Austin,

TX: Acres U.S.A. 2003.

Weston-Allen, D. "In the Interests of Better Health" in *Australian Goat World*, 1981.

Whitby, Coralie. "How do Soils Affect our Diet?" Talk presented to the Orthomolecular Association of Australia. June, 1982.

Widdowson, R. Lecture on soils given at Kiewa Valley Seminars. 1982.

Yiamouyiannis, John. *Fluoride the Aging Factor: How to Recognize and Avoid the Devastating Effects of Fluoride*. Delaware, OH: Health Action Press, 1993.

Index

1080, 268
A Manual of Angora Goat Raising, 3
A Modern Herbal, 185
abscess, 207-209
acclimatization, 46-47
acetonemia, 159, 209-210
Acres U.S.A., 19, 161, 175, 195, 286
acupuncture, 185
aeration, 106-107
afterbirth, retained, 274-275; scabby mouth, 275
age, 48
Albrecht, William A., 288
aloe vera, 185-186
American Sheep Breeders Association, 3
Anderson, Mark, 333
anemia, 210-211
Angoras, 3, 6, 63, 70-72, 299; fleece of, 7-8; and fly strike, 239; Glenroy, 255; kidding areas for, 147
antibiotic, in garlic, 190
antidotes, 110-111
arnica montana, 187-188
arsenic, 264-265
arthritis, 159, 211-213; infective/septic, 212-213

artificial insemination, 39
ascorbate, 181-183
ascorbic acid, 181-183
Auerbach, Charlotte, 171
Australian Goat World, 144
avitaminosis, 213
avocado foliage, and mastitis, 257

Bach Flower Remedies, 185
back legs, 56-57
barber's pole worm, 292-293
barley, 94
bent leg, 213-216
beta carotene, 178
beta-mannosidosis, 216-217
birth, 132-133
black mastitis, 256-257
blackleg, 183, 217-218
bloat, 219-220
blood, in milk, 220
Boers, 10, 66, 74-75; for meat, 337-338
boils, *see abscess*
bonding, 116-117
bone growth, 159
bones, broken, 221-223
boron, 154, 155; and arthritis, 211
bottle jaw, 221

breeding, 37-29, 299-312; and conformation, 301; cycles of, 304-305; and heritabiltiy, 301-303; for long lactation, 300-301; meat goats, 337-339; object of, 299-304; and recessives, 303-304
breeds, 67-88
British Alpines, 3, 80-83, 136, 300
brown stomach worm, 293-294
BSE (Bovine Spongiform Encephalitis), 97
B.T.Z., 273-274
buck runs, 22, 128
bucks, and CAE, 202; management of, 127-132; psychology of, 119-120
Butazoladin, 273-274

CAE (Caprine Arthritis Encephalitis), xi, 45-46, 47, 117, 134, 142, 164, 193-202, 211, 249, 262, 328; and CLA, 226; and milk, 326
calcium, 14, 15, 39, 154, 156-157, 169; and milk fever, 259
calcium ascorbate, 181-183
calcium carbonate, 156
calcium fluoride, 270-271
calcium pantothenate, 180
cancer, 223-224; of skin, 275-276
Capretto, 340-341
casein, in milk, 224
Cashgoras, 9-10, 73-74
Cashmeres, 3, 6, 64-65, 72-73, 300; fleece of, 8-9
castration, of kids, 139-140
cheesy gland, 225-227
chilled animals, 228-229
Cholecalciferol, 183
cider vinegar, 94, 171, 174, 186-187, 234, 312; and arthritis, 211; and bent leg, 214; and casein, 224; and dermatitis, 233; and pneumonia, 264; as potassium source, 19; and udder health, 258; and urinary calculi, 283-284
circling disease, 229
CLA (Caseous lymphadenitis), 126, 147, 193, 225-227
clinical mastitis, 257
clipping, 124
clostridium chauvoei, 217-218
Clostridium perfringens D, 105
cobalt, 160-161
coccidiosis, 227-228
cod liver oil, 276; and bent leg, 214-215; and pneumonia, 264
collars, 126-127
colostrum, 134; and CAE spread, 195
comfrey, 188-189; and broken bones, 222
commercial set up, 315-318
conformation, 48-63
conjunctivitis, 261-262
copper, 16, 17, 91, 154, 161-165; and arthritis, 211; and CAE, 194-198; and cancer, 223; deficiency and diarrhea, 233; deficiency diseases, 162-163; and dermatitis, 232; and foot rot, 241; and Johnes disease, 249; and mastitis, 255; and milk quality, 330; and pox, 230; and worms, 285, 287-288
copper carbonate, 165
copper sulfate, 154; and fluke, 252-253; as vermifuge, 288-289; and worms, 296
cortisone, 272
Corynebacterium pseudotuberculosis, 225
Corynebacteria, 212
coughing, 228
cow hocks, 230
cow pox, 230-231
crossbreds, 87-88

CSIRO (Commonwealth Scientific Industry Research Organization), 15, 174
cyanocobalamin, 180-181
cysteine, 174
cystic ovaries, 245-246

dairy characteristics, 54
dairy goats, nutritional needs of, 93-95
dandruff, 231
de Bairacli Levy, Juliette, 185, 192, 235
de-scenting, 131
defects, hereditary, 243-247
deformities, 231-232
dermatitis, 232-233
diarrhea, 233-234
dicalcium phosphate, 156
disbudding, 138-139
disease check list, 202-203
does, management of, 150-152
dolomite, 14, 15, 19, 52, 105, 107, 110, 154, 156, 157, 158, 159, 179; and arthritis, 211; and avitaminosis, 213; and bent leg, 214; and broken bones, 222; and casein, 224; and diarrhea, 233; and enterotoxemia, 237; and mastitis, 254; and milk quality, 330; and warts, 284
drenches, 284-285; chemical, 291; old-fashioned, 288
drenching, 289-290; strategic, 291-292
driving goats, 356-357
drug reactions, 271-274
dystokia, and potassium, 171

ear tags, 127
early kids, 134
edema, 234-235
ELISA test, 198
Empty Harvest, 333
emu oil, 190

encephalitis, 235-236
enterotoxemia, 125-126, 217, 233, 236-239
Epsom salt, and founder, 242

feed, amount required, 101-102; for bucks, 129-130; during drought, 103-105; for kids, 144-147
feeding practices, 89-113; prior to kidding, 92
feedlots, 27-28
feet, 54, 125
fences, 20-24, 129; electric, 22-24
feral goats, for meat, 338-339
fertilizer, chemical, 14
fireweed, 265
fish meal, 167
flag, 238-239
fleece, 63-66
flooring, 27
fluke, 252-253
fluoride, 268-269
fly strike, 239-240
foot and mouth disease, 240-241
foot rot, 241-242
foot scald, 242
founder, 159, 242-243
French Alpines, 3, 80-83

garlic, 190-191; as antibiotic, 190; and pneumonia, 264
gates, 128-129
goat pox, 230-231
Goat Husbandry, 17, 115, 126, 178, 236
Goat Keeping in the Tropics, 82
Goat Milk Magic: One of Life's Great Healing Foods, 333
goatlings, 147-150; overweight, 55-56
Goats, 39
goats, curing the skin of, 344; for driving, 356-357; for fiber; free range, 118; and fright, 120; history of, 2-11; for hobby, 42,

Index   367

354-357; housing of, 24-28; for meat, 43-44, 337-344; for milk, 313-336; mineral requirements of, 13, 15-16; preparing for show, 348-351; psychological needs of, 115-120; shooting, 342; showing, 345-357; skinning, 343-344; stocking of, 17; for stud, 42-43
goitrogenic feeds, 95
grass tetany, 281
gypsum, 14, 15, 19

hand milking, 318-321
hay, 99-100
Heidi, 335
heliotrope, 107
herbal remedies, 185-192
Herbal Handbook for Farm and Stable, 185, 235
herd testing, 330-332
herding, 41
herpes, 180, 230
Hoard's Dairyman, 144
homeopathy, 185
hormones, 37-39, 272-273
horns, 65-66; in kids, 137
housing, for goats, 24-28
Hungerford's Diseases of Livestock, 164, 217, 233, 287 humus, 15
husbandry, computerized, 37; free range, 33; indoor, 33-37; mountain, 40-41; types of, 33-44
hydatid, 296
hydrogen peroxide, and mastitis, 256

Illustrated Standard of the Dairy Goat, 75
immune system disease, 194
immunizations, 248
impaction, 247-248
injury, 248-249

intersexes, 247
iodine, 19, 165-166
iron, 167; in parsley, 191
irrigation, 19
Ivermectin, 286

jaw abnormalities, 51-52
Jensen, Bernard, 333
John Hunter, 2
Johne's disease, 45, 47, 195, 203, 249-250

kemp, 63-64
keratin, 174
Kervran, Louis, 174
kidding, 308-312
kids, and bent leg, 213-215; and CAE, 197-201; catching of, 198-199; disbudding, 138-139; feeding of, 144-147, 199-201; hand-rearing, 117-118; identification of, 143; management of, 132-150; starting off, 134-135; unthrifty, 247
Kinsey, Neal, 14, 17, 107, 157, 162, 168, 170

La Manchas, 86
lactation tetany, 280
lactations, extended, 39-40
laminitis, 242; also *see founder*
land, choosing for goats, 28-29
leaders, 115-116
leading, 130-131; and kids, 143
leukemia, 250-251
lice, 251-252
lick, 154-155; recipe for, 90
lights, to signal seasons, 38
lime, 14
liver fluke, 252-253
lung damage, 253-254
lungworm, 294-295, also *see worms*

Mackay, Maura, 255

Mackenzie, David, 17, 115, 117, 120, 126, 178, 181, 203, 206, 236, 249
Mad Cow Syndrome, 97
magnesium, 14, 15, 39, 153, 154, 157-160, 169; and milk fever, 259; and tetany, 280
mallow, for coccidiosis, 227
management, 121-152
manure, 15, 36
mastitis, 157, 159, 254-258; procedures for care, 257-258; varieties of, 256
mating, 131-132, 305-307
meat, for home consumption, 341-344; from kids, 340-341
meat industry, 339-241
medicine chest, suggested supplies for, 296-298
medullated fiber, 64
menadione, 184
metaldehyde, 267-268
metritis, 258-259
microchips, 127
milk, and CAE spread, 195; for cheese, 329-330; cooling of, 324; health benefits of, 333-336; heat treating, 200-201; marketing, 328-329; off flavors in, 325-326; for orphaned animals, 328; products of, 328
milk fever, 259
milk goats, buying, 313-314; management of, 314-323
milk replacers, 146-147
milking, 10-11, 323-330; equipment for, 324-325; machines, and CAE infection, 196; small sheds, 322-323
mineral bullets, 155
minerals, 14, 153-176; availability of, 89-90
mismarks, 136
mistletoe, 191
mohair production, 9-10

molybdenum, 167-168
mouths, 136
mycoplasma pneumonia, 118, 262-263
mycoplasmosis, 262-263

nasal bots, 260
natural remedies, 185-192
nematodes, 295
nitrate poisoning, 105-106, 266
nitrates, 169
nitrogen, 168-169
Nubians, 3, 83-86, 136, 300; and wry tail, 247
nutritional requirements of goats, 89-113

Old English Goat, 300
ophthalmia, 261-262
Orf, 275
organic matter, 17
organophosphate poisoning, 266

pangamic acid, 181
pantothenic acid, 180
para-aminobenzoic acid (PABA), 184
parsley, 191-192; for edema, 235
Passwater, Richard, 172
pasteurization, 11, 326-328; and health, 335-336
pasture, for goats, 16-19
peeling horns, 278
pellets, 100-101
pH, 14
phosphates, and Johne's disease, 249
phosphorus, 14, 169, 267
pin worms, 295
pinkeye, 261-262
pleuropneumonia, 262-263
pneumonia, 263-264
poison baits, 267
Poisonous Plants of Australia, 106
poisonous plants, 107-109

Index   369

poisons, 264-269
polled goats, 137
potassium, 19, 167, 171-172; deficiency of, 187
potassium ascorbate, 181-183
predators, 143-144
pregnancy, 307-308
pregnancy toxemia, 269-270
prolapse, 157
prolapse, uterine, 270-271
protein requirements, 96-97
prussic acid, 267
pulpy kidney, 236-239
pyridoxine, 180

Q Star, 238
Q Star tests, 332-333

ration, for dairy goats, 93-94
regulations, for showing goats, 345-246
requirements, for showing goats, 347-248
restraints, 122
retinol, 177-179
roundworms, 295
rugging, 122

Saanens, 3, 75-78, 136
salt, effects of excess, 174
scouring, 233-234
scrota, 61
scrotum, malformation of, 243-244
scurf, 231
seaweed, 154, 165-166; and diarrhea, 234; as feed supplement, 19; as selenium source, 173
seaweed meal, 96, 106, 146, 155, 161; and arthritis, 212; and avitaminosis, 213; and bent leg, 214; and cancer, 223; and foot and mouth disease, 241
selenium, 146, 172-173
Selenium as Food and Medicine: What You Need to Know, 172

silage, 100
skin cancer, 275-276
slippery elm powder, 237
slug bait, 267-268
snakebite, 276-278
sodium, 173-174
sodium ascorbate, 181-183
soil, aeration of, 15, 19; analysis of, 14-15; health and worms, 285-286; top dressing of, 15
spermiostasis, 61-62, 245-246
splinting, 221
split horns, 278
Spyri, Johanna, 335
St. Johns wort, and CAE, 194
stock, acquiring, 45-66
subclinical mastitis, 257
sulfur, 14, 15, 146, 154, 172, 174-175; effects of deficiency, 175; and foot rot, 241; in garlic, 190; and lice, 251
Sundgau, 80
superphosphate, and mineral suppression, 170

tapeworm, 295
tattooing, 140-142
teats, 58-61, 135-136
teeth, 51-53
tetanus, 217, 279-280
tetany, 159, 280-281
tethering goats, 29-31
tetracycline, 118
The Albrecht Papers, 287
The Science of Genetics, 171
thiaminase, 179
thiamine, 179-180
threadworms, 295
thyroid, 165
tick bite, 282
tocopherol, 183-184
Toggenburgs, 3, 78-80, 136
toxoplasmosis, 282-283
trace minerals, 94
travel tetany, 281

treatments, 205-207
trees, as fodder, 16-17, 111-113
tuberculosis, 283

udder, 58; edema of, 234-235; malformation of, 244-245
United Caprine News, 144
unthrifty kids, 133-134
urinary calculi, 283-284
vaccinations, 125-126
VAM (Vitamins, Amino Acids and Minerals), 197, 233; and avitaminosis, 213; and blackleg, 218; and enterotoxemia, 237; and pneumonia, 264; and toxemia, 270
vitamin A, 154, 177-179; and cancer, 223; deficiency of, 38-39; and dermatitis, 233; and cancer, 233; and eye disease, 261-262; and metritis, 258; in parsley, 191; synthesis of, 191
vitamin B complex, 179
vitamin B1, 179-180
vitamin B12, 161, 180-181, 233; and blackleg, 218; and cancer, 223; and CLA, 226; and drugs, 271; and enterotoxemia, 237; and pneumonia, 264
vitamin B15, 181
vitamin B5, 180
vitamin B6, 180
vitamin C, 107, 110-111, 159, 181-183, 188, 212, 233; benefits of, 182; and blackleg, 218; and broken bones, 222; and cancer, 223; and CLA, 226; and dermatitis, 233; and diarrhea, 234; and encephalitis, 235-236; and enterotoxemia, 237; for injury, 248-249; and lung damage, 253-254; and mastitis, 255; and metritis, 258; and pneumonia, 264; and pox, 230; and snakebite, 276-278; and urinary calculi, 283-284
vitamin D, 154, 183; and dermatitis, 233; and metritis, 258
vitamin E, 167, 173, 183-184; and broken bones, 222; and cancer, 223; and dermatitis, 233; for injury, 248-249; and lung damage, 253-254; and pneumonia, 265
vitamin H, 184
vitamin K, 184
vitamins, 177-184
vitamins A and D, and arthritis, 211; and bent leg, 214-215
Vitec, 234, 238, 270
Voisin, André, 162

Walters, Charles, 19
warts, 159, 284
weeds, 16, 98
Weeds: Control Without Poisons, 19
Willis, Harold, 158, 351
worm counts, 292
wormers, natural, 290-291
worms, 284-296; symptoms of, 287; varieties of, 292-296
wry face, 53, 246
wry tail, 51, 247

zinc, 164, 175-176; in seaweed, 176

Acres U.S.A. — books are just the beginning!

Farmers and gardeners around the world are learning to grow bountiful crops profitably — without risking their own health and destroying the fertility of the soil. *Acres U.S.A.* can show you how. If you want to be on the cutting edge of organic and sustainable growing technologies, techniques, markets, news, analysis and trends, look to *Acres U.S.A.* For over 40 years, we've been the independent voice for eco-agriculture. Each monthly issue is packed with practical, hands-on information you can put to work on your farm, bringing solutions to your most pressing problems. Get the advice consultants charge thousands for ...

- Fertility management
- Non-chemical weed & insect control
- Specialty crops & marketing
- Grazing, composting & natural veterinary care
- Soil's link to human & animal health

For a free sample copy or to subscribe, visit us online at
www.acresusa.com
or call toll-free in the U.S. and Canada
1-800-355-5313
Outside U.S. & Canada call 512-892-4400
fax 512-892-4448 • info@acresusa.com